Stepping Out

THREE CENTURIES OF SHOES

by **Louise Mitchell**

with Lindie Ward

FOREWORD BY JUNE SWANN

NEW EDITION

powerhouse publishing
part of the Powerhouse Museum

Acknowledgments

First published 1997
Second printing 1999
New edition 2008

Powerhouse Publishing, Sydney
PO Box K346 Haymarket
NSW 1238 Australia

Powerhouse Publishing is part of the Museum of Applied Arts and Sciences
www.powerhousemuseum.com/publications

Project management: Julie Donaldson* & Judith Matheson*
Design: Kathryn Bird & Ricardo Felipe, Take Me To Your Reader
First edition design: Studio Naar
Editing: Sue Wagner Books & Tracy Goulding*
Photography: Sue Stafford* (main photographer), Scott Donkin*
& Marinco Kojdanovski* (additional photography)
Image scanning: Jean-Francois Lanzarone*
Rights and permissions: Iwona Hetherington*
Deep etching: Andy Chong, PictureThis
Illustrations: Wendy Bishop
Prepress: Spitting Image
Printed by: Phoenix Offset
* Powerhouse Museum

First published in conjunction with the exhibition *Stepping out: three centuries of shoes* at the Powerhouse Museum from November 1997 to May 1998, sponsored by Bally Australia and with assistance from Alitalia.

National Library of Australia Cataloguing-in-Publication
Mitchell, Louise, 1961–
Stepping out: three centuries of shoes.
New ed.
ISBN 978 1 86317 124 3.
Bibliography.
1. Shoes – History – Exhibitions.
2. Footwear Industry – Australia – History.
I. Ward, Lindie. II. Powerhouse Museum. III. Title.
391.413

I would like to thank Jennifer Sanders, deputy director at the Powerhouse Museum, for her support of the *Stepping out: three centuries of shoes* exhibition and publication. I would also like to thank Museum staff who have worked on this project, and I am, in particular, grateful for the contribution of my curatorial colleagues Lindie Ward and Christina Sumner.

A number of people outside the Museum have contributed to the project. Foremost is June Swann, former Keeper of the Boot and Shoe Collection at Northampton Museum, England, who accepted our invitation to catalogue the Joseph Box collection in 1993. I gratefully acknowledge her expertise which she has generously shared over the years. I also thank the British Council for providing a grant which enabled her to travel to Sydney. Another key person is Wendy Bishop who worked as a volunteer researching the shoe collection. Apart from assisting with the documentation of the collection, Wendy's expertise as a shoemaker has provided valuable insight into the craft of shoemaking.

I would like to thank the staff of Museo Salvatore Ferragamo; Musée International de la Chaussure, Romans, France; Musée de la Mode et du Textile, Paris; Victoria and Albert Museum, London; Museum of London; the Bally Shoe Museum of Schönenwerd, Switzerland; the Central Museum, Northampton, England; and the Australian National Gallery, for their assistance.

I am grateful to Sarina Yates and Barbara Fini Stadler of a.testoni who arranged for me to tour the testoni factory in Bologna and to Mr John Hunter Lobb who showed me the bespoke workshop of John Lobb in London. Other individuals who have been particularly generous with information are Donna-May Bolinger, Leonie Furber and Mr Geoffrey Dixon Box, the grandson of Joseph Box.

Louise Mitchell

Front cover image: Woman's barette boots, Joseph Box, England, about 1896. See description page 46.

Back cover image: Woman's heelless shoes, Mario Brini, Italy, about 1959. Ferragamo introduced the steel-reinforced stiletto heel in the 1950s. In these Italian shoes from about 1959 Mario Brini used a special winged-steel support, which replaced the stiletto heel altogether.

Opposite: Woman's shoes 'Kabuki', Beth Levine, NY, about 1964. See description page 66.

Page 2: Woman's button boots, prize work, maker unknown, England 1870–1875. In the 1870s shoemakers showcased their finest skills by displaying prize work at international exhibitions and in prestigious shop windows. The number of hand stitches to the inch (25 mm) was highly valued. In this woman's ankle boot it is 24 stitches.

Contents

The Powerhouse Museum

*T*HE *POWERHOUSE MUSEUM* is delighted to present its world renowned shoe collection in this publication. Originally published in 1997 to accompany the exhibition *Stepping out: three centuries of shoes*, this updated edition contains all of the earlier material plus examples of 21st century shoe design and a new chapter on shoes from around the world.

From the Museum's earliest days, shoes have been acquired by gift and by purchase, although the rationale for their acquisition into the collection has varied over time. To the Museum's first curator, the botanist Joseph H Maiden, shoes were examples of objects 'of all materials of economic value belonging to the animal, vegetable and mineral kingdoms from the raw material ... to the finished article ready for use'. In 1883, the Museum purchased a pair of Cashmir [sic] slippers and a year later a pair of slippers from Turkey. These were acquired more as curios as the costume collection was still very unformed.

During the 1890s, the Museum acquired a group of shoes made from various vegetable fibres and woods, including rice straw bath slippers from Japan and Chinese fibre shoes of *Tilia sp* (linden). Two pairs of Japanese *geta* (wooden platform shoes) made from *Paulownia imperialis* (foxglove) and *Tilia condata* were transferred as a gift from the Royal Botanic Gardens, Kew in London. From these humble beginnings a great collection grew, significantly augmented in 1942 by the purchase of the large and remarkable Joseph Box collection, and in the last 25 years through the selective acquisition of shoes by key designers and manufacturers, both Australian and international.

With the revival of hand-crafted shoes in the last two decades, the Museum's collection has become a source of inspiration for Australian designers and, increasingly, has the capacity to provide access to the great world tradition of shoemaking. Information relating to the collection not only documents and illustrates shoe design and related technologies but also a myriad of stories and histories: the rise and fall of specialist trades, the influence of popular culture, the contemporary revival of a traditional craft, and the fundamental functionality of the shoe in all its diverse forms.

I wish to acknowledge the many staff who have contributed to the revised edition of this popular publication, in particular the Museum's former curator Louise Mitchell and current curators Lindie Ward and Christina Sumner. I extend special thanks to our guest researcher for the first edition, June Swann of Northampton, England, who is enormously knowledgeable about shoes and confirmed for us the significance of this collection. My thanks go also to all others who have contributed to *Stepping out: three centuries of shoes*.

Dr Dawn Casey PSM FAHA
Director, Powerhouse Museum

Opposite: One of the strengths of the Museum's large collection of shoes is the European shoes of the 1700s and 1800s, purchased in 1942 from London shoemakers Joseph Box Ltd. When not on display, this fragile collection is housed in purpose-built storage.

Above: Heeled shoes came back into fashion in the 1860s, and the most popular style was the slip-on shoe known as the court. These silk court shoes were handmade in about 1885 by Henry Marshall, a London bespoke shoemaker. The curved construction of the new heel was a revival of a style from the previous century and was christened the louis.

Foreword

I AM DELIGHTED to write a foreword to a book about shoes at the Powerhouse Museum, for my interest in the collection housed there goes back many years. I grew up in Northampton, England, a town famous for shoemaking for over two hundred years, and worked in the town's museum with the largest collection of historic shoes in the world. Many of them were acquired in the 1890s, which (like our own decade) showed enormous interest in shoes and their history. Amongst those acquisitions were ladies' boots and shoes distinctive for their sculptural quality and fine workmanship, made in the 1880s by the fashionable London bootmaker, Joseph Box.

As I researched our shoes, I discovered in the 1960s that the greater part of the Box collection had been sent to Sydney about 1940, when England had matters more vital to our survival on its mind. I corresponded with staff at the Museum of Applied Arts and Sciences, as it then was, and received a catalogue, which began with a shoe dated about 1550 found concealed in a wall in a house in Chester — a strange superstition which I had been studying for years. The dates and descriptions continued equally exciting, ranging from Ancient Roman to the twentieth century; and not just fashion shoes, but 'prize' work done by the great dons of the nineteenth century, Devlin, Pattison and Player, as well as Joseph Box and his father Robert.

We corresponded for a year or two, and eventually staff moved on, and it seemed impossible to study the collection further at such a great distance, though I continued to ponder and to hope. So I was highly delighted to receive an invitation from the Powerhouse and a British Council grant, which enabled me to visit in 1993, and finally to handle and catalogue the Box collection. Such is the quality that it came not just up to expectations, but exceeded what I had interpreted from descriptions and photographs. Amongst the original wearers were royalty, from Queen Victoria and Edward VII to the Duchess of York, notorious in 1791 for her tiny feet. There were ordinary working shoes not just from Britain, but from those other countries the English had explored, adults' and children's, so evocative of the past and faraway places. Truly, one of the world's great shoe collections, though sadly, I feel, more relevant to English history than Australian.

Shoe students always say that you can tell the history of the world from its footwear, while others are always surprised at the popularity of shoes, a subject men and women expound on at great length with the slightest encouragement. Perhaps a little of my enthusiasm has remained, and at last many of the Powerhouse shoes will be displayed, to amaze and amuse, and to inspire modern shoemakers and designers whose works, I am delighted to hear, are also represented; indeed to appreciate the most sensitive trade in the world, 'the Gentle Craft', always responsive to human frailties. So if you are tempted to laugh at the strange shapes, analyse your own shoes first: they will look equally bizarre and unsuitable for human wear in a few years' time.

June Swann, 1997
Former Keeper of the Boot and Shoe Collection at Northampton Museum, England

Introduction

*F*ROM ITS BEGINNINGS in the aftermath of Sydney's International Exhibition in 1879, the Powerhouse Museum has had a broad collecting scope. In its early years a diverse range of artefacts was gathered to provide a collection based on industry and commerce which would serve an educational role. The proximity of a shoemaking course at the neighbouring Sydney Technical College led to the collecting and exhibiting of shoes that were intended to be instructive and inspiring for college staff and students.

The idea of acquiring an existing historical footwear collection came originally from the staff at the college's Boot and Shoe School who felt that an historical perspective would enhance the existing display of modern shoemaking. In 1939 Arthur Penfold, then director of the Museum, was in London seeking exhibits that would be 'of interest to Australians'; at the suggestion of the editor of a trade journal, the *Footwear Organiser*, contact was made with a shoemaking company called Joseph Box Ltd about a collection amassed by the present owner's uncle and grandfather during the second half of the 1800s.

Writing back to the Museum, Penfold described Joseph Box Ltd's collection as having about 300 shoes including 'hand made shoes from 1500 to date ... The stitching is so fine that a magnifying glass is required to see some of it'. So began a lengthy negotiation which concluded in 1942 when the Museum purchased the Joseph Box collection. The pressures of World War II presumably added to the urgency of finding a safe home for the collection as the company itself was winding down and was taken over by bespoke (made-to-measure) shoemakers John Lobb Ltd some time after 1953.

Joseph Box Ltd had its origins in a London shoemaking business established in 1808 by a 'ladies shoemaker' called James Sly. From 1816 Sly's apprentice was Robert Dixon Box, the fifteen-year-old son of a bankrupted Quaker attorney. When Sly moved his business to 187 Regent Street in 1824 he made Robert his right-hand man. Sober, hard working and deeply religious, Robert was appointed manager of the business by Sly's executors when Sly died in 1826. The promotion was made over Sly's son, who apparently was not considered suitable, and when in 1832 Robert became owner of the business the younger James Sly was apprenticed to him. This situation led to a certain amount of friction, as Box family records claim that Sly tried to stab his master with a clicking knife and had to be restrained by workshop employees.

Despite such setbacks, under the ownership of Robert the workshop gained a reputation for fine shoemaking through its participation at international exhibitions and by obtaining Royal Warrants.

The business became known as Joseph Box Ltd after it was transferred to Robert's son Joseph in 1862. Like his father, Joseph started in the trade at the age of 15, but retired at the relatively early age of 42 to enable his daughters to enter society. Although he transferred the business to his cousins the Box Kinghams in 1882,

Above: The core of the Powerhouse Museum's shoe collection was amassed in the second half of the 1800s and was purchased from Joseph Box Ltd, a London shoemaking business. Pictured is Robert Dixon Box (1801–1880), father of Joseph Box, who began his apprenticeship in shoemaking in 1816 and purchased his master's business in 1832.

Oil painting by J R Dicksee (1817–1905). Courtesy Geoffrey Dixon Box

Opposite: This photograph taken in the late 1800s shows the shop of Joseph Box Ltd at 187 Regent Street, London. The sign 'Court Boot Maker' refers to the company's Royal Warrants of Appointment: the Princess of Wales, the Duke of Edinburgh and the Crown Princess of Germany. By 1899 the business shared premises with Gundry & Sons, Queen Victoria's shoemaker.

Courtesy City of Westminster Archive and Geremy Butler

Above: Joseph Box (1840–1922) was the son of Robert Dixon Box and took over his father's business in 1866. He retired at the age of 42 but maintained an interest in shoemaking through collecting and organising exhibitions of historical shoes.
Courtesy Geoffrey Dixon Box

Joseph maintained an active interest in shoemaking through collecting. In 1889 he organised a widely publicised exhibition at 187 Regent Street of his own collection and others with the aim of showing 'the progress made in the art of shoemaking, especially in the finer and more expensive kinds worn by the wealthy classes'.

The collection acquired by the Museum was probably started by Robert Dixon Box and consolidated by Joseph Box and the Box Kinghams during the second half of the 1800s. It includes remnants of leather shoes from the Middle Ages found in English archaeological sites, intact European shoes from the 1600s onwards, 'foreign' shoes collected as 'curiosities' from around the world, shoe buckles and spurs, as well as documents relating to Joseph Box Ltd.

A selection of the collection was on display in the old Museum of Applied Arts and Sciences from 1954 to 1978 but when this building closed and the development of the Powerhouse began, the collection went into storage. The new Museum opened in 1988, and while shoes have featured in various exhibitions, the main shoe collection, of which the Joseph Box collection is the core, has remained in storage. Conscious of its significance, the Museum invited footwear scholar June Swann, former Keeper of the Boot and Shoe Collection at Northampton Museum, to come to Sydney and catalogue the collection. As dates and descriptions of shoes were amended and provenances clarified, the collection was 'rediscovered' and the time seemed right to display it again. The temporary exhibition *Stepping out: three centuries of shoes* provided an opportunity to view a collection that had not been seen for many years. This publication presents a selection of the shoes from the exhibition.

Since the purchase of the Joseph Box collection, the Museum has continued to collect shoes. The Museum's collecting scope is broad and covers decorative arts, design, social history, science and technology. Shoes are collected as much for their historical associations as for their aesthetic and technical qualities. Cricket boots signed by the famous cricketer Don Bradman and rubber boots worn by an Antarctic explorer, Dr Peter Towson, are examples of footwear acquired primarily for their Australian provenance. Shoes that represent styles of dress which evolved independently of European fashion are also collected, either as examples of good or significant design or because they have interesting technical features.

However, the majority of shoes acquired since the purchase of the Joseph Box collection have complemented it by representing mainstream fashions in shoes, and in particular stylistic changes in the 20th and early 21st centuries. It is the development since 1700 of the fashionable shoe, accompanied, and in some cases made possible, by technical development, which is the main focus of this book. There is also a chapter on traditional shoes from around the world showing the great diversity of styles from different regions. The final chapter presents aspects of the history of Australian shoemaking.

Above: Sydney Technical College in Ultimo established a shoemaking school at Erskineville in 1906. The school was relocated to Ultimo in 1939 and its proximity to the Museum encouraged the curator to collect and display historical shoes. This photograph of about 1919 shows a shoemaking class conducted by the college under the commonwealth government's repatriation training scheme.
Courtesy Sydney Technical College

Left: Archival material relating to the Joseph Box collection includes scrapbooks compiled by Joseph Box in the late 1800s and early 1900s.
Courtesy Geoffrey Dixon Box

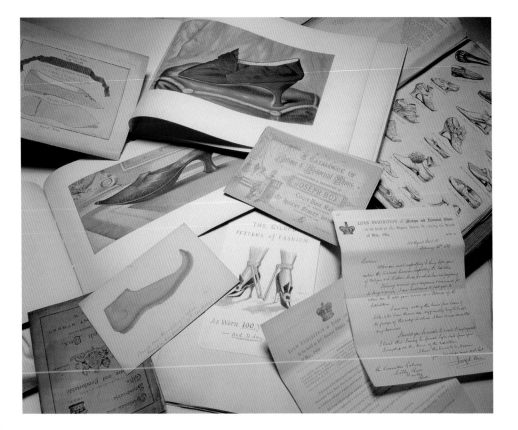

The making of a shoe

Above: These shoemaking tools were part of the Joseph Box collection and date from the 1800s. The lasting pincers (left) were used to pull the upper leather tight around the last before it was secured with nails. The moon knife (right) was used to cut leather.

LEATHER IS USUALLY associated with shoemaking, although wood and plant fibre are also important materials in the history of the craft, particularly in societies with strong traditions of basketry and woodcarving. It is leather, however, that has become the predominant material for making shoes, favoured as it is for its flexibility, practicality and comfort.

The first step in constructing a leather shoe is the making of a last, a smoothly contoured and stylised wooden or moulded-plastic model of the foot. The last determines the shape of the shoe. The production of the last can be a time-consuming and expensive process as a different pair of lasts is required for each shoe style, whether the shoes are handmade or mass produced.

Using the last as a guide, the next step is to cut out the leather for the sole, the insole and the upper. The sole is a shoe's bottom or ground contact piece of material. The insole is the inside bottom part of a shoe on which the foot rests; the upper covers the top of the foot and usually consists of a number of pieces including the vamp, the front section of the shoe covering the toes and part of the foot's instep, and the quarters which cover the upper sides of the foot. The cutting of leather requires considerable skill as the cutter or 'clicker' must take into consideration the elasticity of the leather and the direction of stretch, as well as aesthetic qualities such as grain and colour, and the most economical use of the piece of leather.

After the cutting, the 'closer' joins the pieces of the upper and incorporates the lining. Then the leather is fitted precisely to the last, first the insole and sole and then the upper, through the skilful combination of pulling, stretching, shaping and temporary nailing. The sole is then stitched on, and the heel constructed in wood, plastic or layers of leather.

Modern shoe construction was widely adopted by the 1500s, superseding the turnshoe construction technique, characteristic of the Medieval period, where the shoe is constructed inside out and then reversed – or turned – so that the seams are on the inside. The turnshoe technique is still used for specialty footwear such as ballet shoes and slippers. An important innovation in Medieval shoemaking was the invention of the 'rand', a wedge-shaped strip of leather sewn between the upper and the sole to make the shoe more waterproof. This developed into the welt construction method where a strip of leather wider than a rand is stitched onto the edge of the upper and insole and the sole is then attached to the welt. The welted construction technique was used for making heavy protective boots from the early 1500s when the use of heavier soles and thicker leather made it impossible to turn a shoe inside out as in the turnshoe construction technique.

Left: Donna-May Bolinger is a Sydney-based bespoke shoemaker and designer who employs the same basic tools and techniques used by shoemakers for hundreds of years. She is pictured here with the tools of her trade including boot lasts and patterns, a hammer and lasting pincers.
Photo: Eddie Ming

Whether undertaken by a shoemaker working alone or in a workshop with artisans performing the specialist tasks of the 'clicker', the 'closer' and the 'maker', the process of shoemaking remained unchanged from the 1500s until the mid 1800s when mechanisation took over from handwork. Although most shoes today are made in factories with sophisticated machinery, and modern methods of attachment use a variety of cements and adhesives, there are still bespoke establishments catering for those who appreciate – and can afford – the robustness, flexibility and comfort of the handmade shoe.

Traditional shoes from around the world

*S*ANDALS, CLOGS, PUMPS, slippers, boots, thongs, platforms – types of shoes that have come and gone in Western fashion over the years – are styles that were often influenced by traditional shoe designs from other regions. The great diversity of materials, techniques and forms seen in shoes from around the world can tell us about the wearer's profession, religion, national and regional identity, class and gender. Sometimes cultures separated by vast distances have developed similar solutions to problems of design and construction while others are unique.

Sandals are an ancient form of crafted shoe and are worn throughout the world particularly in hot climates. Sandals made from rawhide suit desert conditions and were traditionally worn in the Middle East. In Africa where most people went barefoot, toe-thong leather sandals were a status symbol and worn by royalty. Likewise in the Indian continent, sandals have become associated with special occasions and high status. Sandals made of woven plant fibre were also worn in China, Japan and the Pacific Islands.

In India, where Hindu beliefs forbid the use of cowhide, toe-knob sandals were made of decorated wood, ivory or silver. Some Indians also wore sandals that featured a sole raised on two stilts for visiting temples. In the Middle East women wore elevated sandals for visits to the bath house to help keep their feet dry. In Japanese cities both men and women wore the elevated sandal known as the *geta*.

The custom of removing shoes to go indoors is common to many cultures. Slippers with backs folded down, for easy removal before entering a mosque, are seen throughout the Middle East. In North America, indigenous people developed a type of flexible slipper from leather which became known as the moccasin. After contact with traders, they decorated moccasins with glass beads from Europe.

Boots were commonly used for horse riding in many regions. European shoemakers admired the elaborately made and decorated horse-riding boots from the Middle East. Different communities made use of materials at hand. In the desert, boots were made of rawhide, whereas in snow regions, they were made from seal or reindeer skin. Slippers for indoor wear could be worn inside leather boots for extra warmth.

For many centuries Europeans have worn wooden shoes such as clogs. Worn with socks for warmth, clogs were practical in muddy conditions. Chinese labourers also wore shoes made of wood and vegetable fibres. Perhaps the most intriguing shoes from China are those made from silk with brightly coloured embroidery of auspicious symbols. These include the elevated shoes worn by Manchurians as well as the tiny 'lotus' shoes worn in premodern China by women with bound feet.

The shoes in this chapter date from 1830 to 1900. They were part of the collection of 19th century London shoemaker Joseph Box and many featured in the 'foreign' sections of his exhibition *Antique and historical shoes.*

Opposite: These shoes from the Joseph Box collection dating from 1830 to 1900 come from around the world and are evidence of diverse traditions in shoemaking. Many of them featured in the 'foreign' section of the exhibition entitled *Antique and historical shoes,* which Joseph Box organised at 187 Regent Street, London, in 1889.

Man or woman's waraji or sandal

JAPAN, 1850–1900

Simple shoes like these were made wherever people plaited or twisted plant fibres together to form string. Similar sandals have been uncovered in archaeological sites in such widely separated regions as South America, Egypt and Central Asia and they are still made and worn in the Pacific Islands, Spain and Japan. Originally thought to be African, these sandals are now identified as Japanese *waraji*.

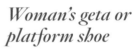

Above: Detail from *Three kabuki actors* by Kuniyoshi Utagawa, 1830–1850s, Japan. Colour woodcut, 357 x 240 mm.

Bequest of Kenneth Myer, 1993. Collection Art Gallery of New South Wales

Below: *Rain of the fifth month* from the series *Customs of the four seasons in the poetry masters* by Suzuki Harunobu, late 1760s–1770, Japan. Colour woodcut, 273 x 220 mm.

Purchased 1951. Collection Art Gallery of New South Wales

Woman's geta or platform shoe

JAPAN, 1850–1900

Geta are wooden platform shoes worn outdoors in Japan and removed on entering a house. By elevating their feet from the ground, wearers kept long kimonos or *yukatas* clean. Special socks called *tabi*, with a division between big toe and second toe to accommodate the V-shaped thong, were sometimes worn underneath. The style and finish of *geta* varied, signalling the gender and status of the wearer.

Boy's slip-on cloth shoe

CHINA, 1850–1895

Chinese women traditionally made cloth shoes for their families, choosing sober colours for adults and bright colours for children in order to scare away the evil spirits. The soles were constructed from recycled fabrics, pasted and stitched in layers to make a firm platform, and attached to the uppers by professional shoemakers who travelled around the country from job to job.

Above: A shoemaker, tinted engraving from *Views of 18th century China* by William Alexander and George Mason.
Studio Editions

Below: A Mandarin in his summer dress, tinted engraving from *Views of 18th century China* by William Alexander and George Mason.
Studio Editions

Man's slip-on basket-woven shoe

CHINA, 1850–1895

This type of slip-on mule made from local plant fibres was commonly worn in rural areas of southern China in the late 19th century. The open basket-weave structure lets the feet breathe and suits the humid summers. Chinese officials often wore mules like these with white cotton socks. Exhibited in England in 1895 and 1897, the original Box collection label described them as 'African, Niger River'.

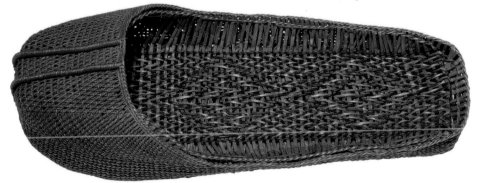

Above: A woman sewing stockings, tinted engraving from *Views of 18th century China* by William Alexander and George Mason.
Studio Editions

Below: A village dweller. Madrasi School, India. Tempera, gouache, gold and silver leaf on paper, 216 x 159 mm.
Gift of Mr George Sandwith, 1957. Collection Art Gallery of New South Wales

Woman's slip-on 'lotus' shoe for a bound foot

CHINA, 1880–1895

A woman's tiny feet, achieved through painful binding from an early age, were considered both sexually attractive and a status symbol in Han China. In addition, making and embroidering their own shoes was regarded as a desirable womanly refinement. This particular 'lotus' shoe, which was exhibited in London in 1897, once belonged to H E Marchioness Tseng, wife of the Chinese Ambassador to the Court of St James in the 1890s.

Man's toe-ring sandal

SOUTHERN INDIA, MID 1800s

Now identified as Indian, these toe-ring sandals were originally described as ancient Roman when exhibited as part of the Box collection in England in 1889. In traditional contexts in India it is common for men to wear shoes, while women go barefoot and decorate their feet with gold and silver jewellery, such as anklets and toe rings. In northern India, women also paint their feet with henna patterns.

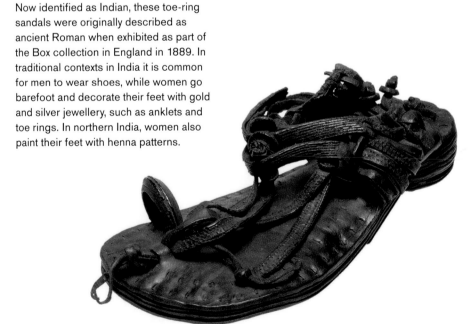

Man's mule

INDIA, LATE 1800s

By folding the heel flat, mules like this could be slipped on and off easily as it was customary to go barefoot indoors. The vamps and underside of the turned-over toes are beautifully decorated with embroidery, beetle wing pieces and shagreen. This style of shoe originated in Persia (now Iran), and was brought to India by the Mughals in the early 1500s. Mules are still widely worn in India and are also popular souvenirs for visitors to the subcontinent.

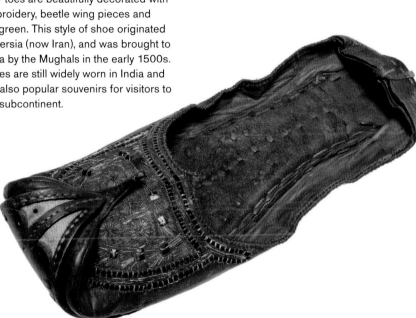

Above: A Hindu of the Carnatic 1769–1772. Madrasi School, India. Tempera, gouache, gold and silver leaf on paper, 216 x 175 mm.

Gift of Mr George Sandwith, 1957. Collection Art Gallery of New South Wales

Below: A Brahmin and his wife. Company School, India, about 1800. Opaque watercolour with gold on paper, 230 x 180 mm.

Gift of Mr George Sandwith, 1957. Collection Art Gallery of New South Wales

Man's ivory toe-peg sandal

INDIA, MID TO LATE 1800s

Sandals with stilts keep the wearers' feet well off the ground. Having no ties or fastenings, these clog sandals are held in place by gripping the peg between the toes, and are easy to slip on and off. They are said to have belonged to a Brahmin priest who wore them to the temple. Brahmin Hindus are specially concerned with ritual purity and pay close attention to cleanliness and the materials they wear.

Above: *Almea rosa* by Alberto Pasini, 1878.
Oil on board.

Private collection. Courtesy Vittoria Botteri Cardoso

Opposite: *Dame franque et sa servante* by
Jean-Etienne Liotard, 1742–43. Pastel on
parchment, 710 x 530 mm.

© Musee d'Art et d'Histoire (Cabinet des Dessins), Ville de
Geneve. Inv no 1936-17. Photo: Bettina Jacot-Descombes

Woman's mule overshoe

TURKEY, MID 1800s

This type of backless protective mule
with its pointed, upturned toe is common
throughout the Muslim world. With a
strong sole, short vamp and upturned toe,
overshoes like this were easily slipped on
over slippers or soft leather boots when
walking outdoors. The inner red felt sole
is decorated with a stylised leaf spray in
the metal-thread embroidery characteristic
of Turkey.

Woman's bath clog

PALESTINE, ABOUT 1890

High-footed clogs were worn by women
in Muslim communities when visiting a
public bathhouse, in order to raise their
feet clear of the wet bathhouse floor. As
women left their homes comparatively
rarely, outings to the bathhouse were
generally considered special occasions.
Consequently their clogs were usually
highly decorative. In this pair, the soles are
inlaid with a geometric design in mother-
of-pearl outlined with silver. The strap or
band to hold the clog in place is missing.

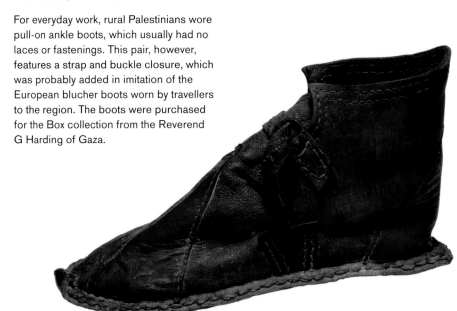

Man's ankle boot

PALESTINE, LATE 1800s

For everyday work, rural Palestinians wore pull-on ankle boots, which usually had no laces or fastenings. This pair, however, features a strap and buckle closure, which was probably added in imitation of the European blucher boots worn by travellers to the region. The boots were purchased for the Box collection from the Reverend G Harding of Gaza.

Above: Shoemaker, Bethlehem market, *1930s.* Black and white photograph by Jan McDonald.
Courtesy Shelagh Weir

Below: Detail of *Horse market, Syria* by Alberto Pasini, 1893. Oil on canvas, 510 x 690 mm.
Collection Art Gallery of New South Wales

Yezmeh or calf-length riding boot

DAMASCUS, SYRIA, 1895–1905

Made using invisible 'tunnel' stitching, these red boots appear seamless and are of a type said to have been favoured by Bedouin sheiks. Tunnel stitching was developed by Arab leather-workers, whose skills were greatly admired and imitated in Europe. 'Cordwainer', the medieval English name for a shoemaker, refers in fact to the Arab city of Cordoba in Spain, which was known for its fine, brightly coloured leather.

Man's sandal

PALESTINE, 1895–1905

Simple sandals like these could be made in about half an hour from unprocessed skin, or rawhide. More quickly produced than most leathers, whose time-consuming tanning involves a number of steps, rawhide is dry and tough like cardboard but becomes soft and limp when wet. The sandals are of the kind traditionally worn by the Bedouin in Palestine.

Above: Bedouin wearing sandals, British Mandate period. Black and white photograph by S I Schweig.
Courtesy Shelagh Weir

Below: West African king. Black and white photograph by T F Garrard.
© abm (Archives Barbier-Mueller), Studio Ferrazzini-Bouchet, Geneva

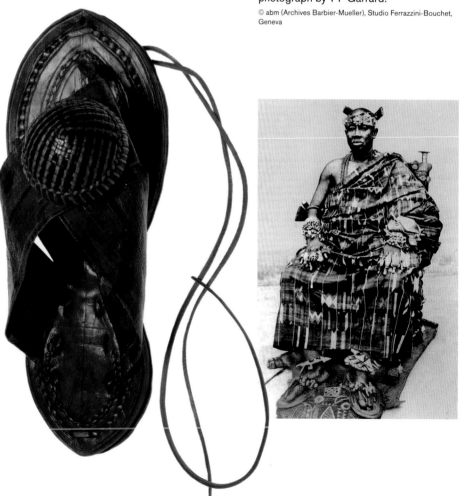

Man's toe-thong sandal

AFRICA, MID TO LATE 1800s

In parts of the world where most people go barefoot of necessity, shoes sometimes become a status symbol and are only worn by people who don't walk much. The West African Hausa king in this photograph may well have had a 'sandal-bearer' who carried spare sandals in case a strap broke and to ensure his royal feet did not touch the ground. These sandals, with their added dome-shaped basket-weave ornaments, were displayed in London in the 1890s.

Above: Chan-Cha-Uia-Teuin, Teton Sioux woman by Karl Bodmer. Watercolour and pencil on paper.

Gift of Enron Art Foundation. Courtesy Joslyn Art Museum, Omaha, Nebraska

Below: The Greenland hunters, Qiperoq and Poq, unknown artist, Copenhagen, 1724.

Cree Indian moccasin

CANADA, MID TO LATE 1800s

This type of soft, flexible shoe is known as a moccasin, traditional everyday footwear for the indigenous peoples of North America. Moccasins were made by the women, who treated animal skins to make soft leather and used coloured glass beads bought from fur traders as decoration. In cold weather, people padded their moccasins with moss for warmth and covered them with overshoes to protect them from the snow.

Child's slip-on shoe

GREENLAND OR ICELAND, 1830–60

Low slip-on shoes like this are worn by the indigenous people of both Greenland and Iceland. The design was probably adapted from the shoes worn by Norse immigrants who settled the area in the Middle Ages. This shoe is made from thin parchment-like sealskin with a fur trim round the edges and delicate butterfly appliqué decoration on the vamp.

Woman's mule clog

NORTH ITALY OR PORTUGAL, LATE 1800s

Because of their practicality in different weather and work conditions, clogs were worn all over Europe, by farmers working in muddy fields, fishermen on wet docks and women in the cobbled streets. Clogs came in a range of shapes, either encasing the whole foot or backless and sometimes with a fabric or leather upper. Today they have largely been replaced for work wear by rubber boots.

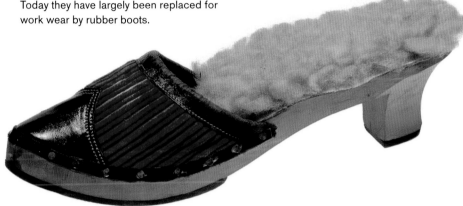

Above: Traditional Portuguese dress, detail from *Le costume historique* by Auguste Racinet, Paris, 1888.

Below: A street in Brittany (Cancale) by Stanhope Forbes, 1881.

Infant's or miniature clog

FRANCE, LATE 1800s

Wooden clogs, like this tiny pair from the Ariège in southwest France, are carved in one piece. Because of their rigidity, upcurled toes are an essential part of the design, enabling the wearer to walk easily by rocking forward on the toes. The durability of clogs ensured they were handed on to younger members of the family, who easily adjusted them to fit by wearing layers of socks.

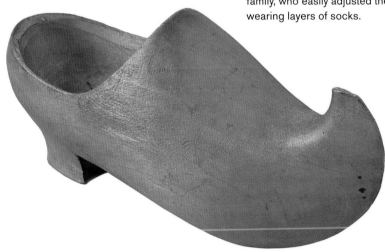

Fashionable footwear: the 18th and 19th centuries

*T*HROUGHOUT THE 1700s and 1800s fashionable footwear reflected social, economic and political developments. Before the French Revolution the style of dress favoured by the French court influenced fashion throughout Europe and England, although few outside Paris could emulate the elaborate styles of the French aristocracy.

The high heels and fancy buckles that were part of the aristocrat's style of dress seemed to disappear overnight in 1789, but in reality the move to a simpler style had already begun. Partly as a response to the interest in the arts of the classical world, women's shoes became heelless and reminiscent of the sandal worn in ancient Rome. Men also wore shoes, tied instead of buckled, but boots were more common. The early 1800s were dominated by the Napoleonic Wars and their aftermath. Shoemakers adapted the ubiquitous military boot and first men, then women too, adopted boots as everyday wear. They remained the principal style of footwear until the late 1800s, when shoes became popular again.

Fashion was also expressed through changing toe shapes, heel heights, materials and decoration. Illustrations in magazines in the late 1700s encouraged the adoption of new styles. The quickening pace of fashion reflected an expansion of manufacturing and the development of mass consumerism. While shoes were still made by hand through the 1700s, there were large workshops producing only ready-made shoes which were sold from the workshop, from warehouses and by pedlars. By the early 1800s, many shoemakers were selling ready-made shoes as well as making shoes to order, using tools and construction methods unchanged for centuries.

Although high levels of skill in the shoemaking craft were seen at the international exhibitions held during the second half of the 1800s, the individual shoemaker virtually disappeared as mechanisation took over process after process. By the end of the century most shoes were made in factories and sold through specialist shoe shops.

Opposite: This miniature mule, shown here in full scale, was made for exhibition and was probably displayed at London's Great Exhibition held in 1851. The exaggerated heel and toe are based on fashion of the 1700s and reflect the Victorian fascination for high-heeled shoes of the type worn by Marie Antoinette and her contempories at the French court.

Before the French Revolution

Above: In this satirical print dating from about 1750, a shoemaker is festooned with the tools of his trade. He wields a measuring stick and holds a hide like a shield, wooden lasts hang from his neck and his clothing is embellished with files, a leather-cutting knife, awls and pincers.
Etching by C Horman, published by M Engelbrecht, Augsburg, Bavaria. Courtesy Bally Shoe Museum

Right: This unassuming child's leather shoe, in style dating to the early 1700s, was found concealed in a house demolished in England early this century. During the 1700s shoes were sometimes used as charms and when a house was being built, a shoe was placed in a wall to ward off evil spirits.

WOMEN'S FASHIONABLE SHOES in the early 1700s had a pointed toe and a sturdy waisted heel; they were fastened with latchet buckles or a tie over the instep. Fashionable shoes had uppers made of patterned silks and linens which did not necessarily match the material of the wearer's dress. More practical shoes were made in wool or leather. To protect these costly shoes from muddy unpaved streets, overshoes known as clogs or pattens were worn. The clog had a leather sole and a covered wooden wedge which fitted under the arch and housed the heel. The patten was similar, but was mounted on an iron ring to lift the wearer higher above the mud.

The shape of women's shoes became more refined as the century progressed. Under the influence of the rococo, a court-based decorative and ornamental style originating in France, women's shoes were made in lighter coloured fabrics. Heels became higher and more slender. The most evocative shoe of the period was the mule, a type of backless slipper, which typified the sexual allure of fashion in the 1700s.

In stark contrast to the types of shoe worn by women, men wore plain dark leather shoes with low heels. An exception was the decorative shoes sometimes worn at court. Early in the century shoes were square-toed, but in the 1740s a rounder toe was preferred. They featured decorative buckles which could be transferred from one pair of shoes to another, and were regarded as a type of jewellery. From the 1720s onwards buckles gradually increased in size and importance as a status symbol. Elaborate buckles for court dress were made of silver and decorated with diamonds, while more basic ones were of metal such as steel. A fashionable man or woman had many pairs of buckles.

Above: The uppers of these women's shoes are made from costly brocaded silk and embroidered linen. The shoe in the foreground has a matching overshoe which fits under the arch of the shoe and protects it from dirt. The shoes were made in England and date from 1700 to the 1740s.

Opposite: High-heeled backless slippers called mules were fashionable throughout the 1700s and complemented the highly decorative and feminine styles of rococo dress. This leather mule, made in France in the 1770s, has been trimmed with silk ribbon.

Above: Chinoiserie, a European interpretation of Chinese style, manifested itself as a brief shoe fashion about 1787. This shoe has the distinctive upturned toe of Chinese footwear. It is made of leather and has been lined with wool, presumably to provide extra warmth.

Opposite: Although this man's shoe was most likely worn to the coronation of George III in 1761, its square toe was a stylistic feature of men's shoes in the early 1700s. It is made of leather and has been decorated with a lace ribbon but is missing its paste buckle.

Opposite below: The elaborate nature of these silver and paste shoe buckles suggests they were worn for special occasions. Buckles were treated like jewellery and were transferred from one pair of shoes to another. These were made in England in the 1780s.

Right: Mounted couriers and postilions wore practical thigh-high boots for protection. Made of thick leather and virtually indestructible, postilion boots like this one dating from the late 1700s became obsolete as roads improved.

Post revolution and the early 1800s

The fancy buckles and high heels that distinguished the aristocracy of the *ancien régime* were out of favour after the French Revolution in 1789. The change towards a simpler 'English style' had begun before the revolution but the events of 1789 accelerated the pace. Inspired by the arts of ancient Greece and Rome, women wore chemise style dresses in plainer fabrics and the high-heeled shoe was replaced with a slip-on leather shoe with a low heel. By the 1820s the heel had completely disappeared and the toe was blunt. These flimsy shoes were often worn with ribbons to cross and tie around the ankle, reminiscent of the classical Roman sandal. Although impractical for outdoor wear, they were suitable for dancing and their design reflected the popularity of balls in social life at the beginning of the 1800s.

For men, the latchet tie replaced the buckle as a fastening for shoes. It was the boot, however, which became the principal form of fashionable footwear for men at the end of the 1700s. The popularity of boots was influenced partly by Englishmen's country wear, which included riding boots, and partly by military dress. Many of the new boot styles were named after military heroes of the era of the Napoleonic Wars. These included the wellington boot, a slim-cut leather boot which was worn under the narrow trousers that superseded breeches; the hessian, named after the German state of Hesse, a knee-high boot cut on a V at the front and sometimes decorated with a tassel; a high-cut boot with an extension over the knee was dubbed the napoleon later in the century; and the blucher, originally an army shoe, was a laced ankle boot.

The demise of the heel meant that shoes could more easily be made for left and right feet, making them more comfortable. It was the difficulty of making mirror-image lasts for heeled shoes, first worn in the 1600s, that had encouraged shoemakers to make 'straights', that is, shoes that fit either foot.

Paris. Cordonnière.

Above: This print, dating to about 1810, depicts a Parisian shoemaker's wife. Her apron contains shoes and she carries a shoehorn. By this time shoemakers were keeping a stock of ready-made shoes, where in earlier times each pair would have been made to order.

Tinted copperplate etching by Lanté and Gatine. Courtesy Bally Shoe Museum

Below: Slip-on shoes made from coloured kid leather suited the simpler style of dress worn by women after the French Revolution. These date from the 1790s to about 1810 and have the characteristic pointed toes and low heels associated with the period. The black shoe has a more durable quality and is typical working class wear of the period.

Opposite: Ankle boots for women, known as high-lows, were first worn as fashionable footwear in the early 1800s. These have silk uppers and were probably worn in England as part of a wedding outfit in 1804.

Above: The tie replaced the buckle as a form of fastening after the French Revolution. These men's leather shoes are tied with silk ribbon. Shoes like these were suitable for dancing and were worn in the drawing room in preference to boots. The square toe dates this pair to the 1830s.

Opposite: Heelless shoes with a squared toe were worn by women between about 1830 and the 1860s. A range of materials, including silk, woven straw and imported kilims, has been used for the upper parts of these shoes. The paper labels indicate they were purchased ready made.

Opposite: Many of the names given to the different styles of boots fashionable in the 1800s show a preoccupation with war. Boots cut high in the front like these were first worn in the 1730s and became known as the napoleon boot in the following century. This pair was made for exhibition by Edward Pattison, a London shoemaker, between 1850–1870.

Above: In this depiction of a French shoemaker's workshop in the early 1800s, a shoemaker measures a customer's foot; he is assisted by an apprentice. In the background a cutter prepares the leather and journeymen sew by the window, while the women engage in the sale of ready-made shoes (unknown illustrator).

Musée Carnavalet, Paris. Photo: Jean-Loup Charmet

The mid 1800s

Above: The balmoral boot, named after Queen Victoria's Scottish estate, was a front-laced ankle boot. It was popularised by Prince Albert who is said to have liked the style for its slenderising effect. This men's balmoral was made for exhibition by London shoemaker William Walsh in the 1850s.

Opposite: The elastic-sided boot was invented by London shoemaker Joseph Sparkes Hall. In 1837 he presented this prototype pair to Queen Victoria. The boot's gussets have cotton-covered metal springs which enable the boot to be pulled on and off with ease. The elastic-sided boot had become the most popular style of footwear by the mid 1800s.

Worn first by men, the boot was the most popular form of footwear for both sexes by the mid 1800s. The differences between men's and women's boots reflected their different roles in society. While men wore sturdy leather boots suitable for an active outdoor life, women wore ankle boots, usually with cloth uppers, under long, full skirts which made movement difficult and were impractical for anything other than an inactive life in a domestic setting.

The difficulties of fastening a boot with buttons and laces led London shoemaker Joseph Hall Sparkes to experiment with fastenings. His prototype version of the elastic-sided boot was presented to Queen Victoria in 1837, the year she came to the throne. In *The book of the feet* Sparkes Hall claims the Queen was well satisfied and that subsequent improvements in his design 'combined to make the elastic-sided boot the most perfect thing of its kind'.

Although the Industrial Revolution had brought mass production in many industries by 1850, shoes were still being made by hand. Technical skill in the craft was clearly evident at London's 1851 Great Exhibition, an important showcase of mid-century style and developments in art and industry from around the world. These exhibitions fostered a spirit of competition among shoemakers and a pride in hand craftwork. Shoemakers submitted 'prize' shoes and boots, hand-lasted and stitched, with an emphasis on materials, tanning and fine, even stitching.

The international exhibitions of the 1850s to 1870s became vehicles for promoting shoemaking and establishing a craftsman's reputation, as illustrated by the story of John Lobb. After training as a bootmaker in London, Lobb tried his luck on the Australian goldfields where, according to his biography, he made hollow-heeled boots for prospectors to hide their gold. He set up an establishment in Sydney in 1858 and from there participated in London's 1862 exhibition; he was the only Australian entry and won a gold medal. The following year he sent a pair of riding boots to the Prince of Wales and was awarded a Royal Warrant. On the strength of this recognition, he returned to London and set up a business 'John Lobb, Bootmaker', which continues today as the world's most famous bespoke (made-to-measure) shoemaking establishment.

Above: Shoemakers exhibiting at London's Great Exhibition in 1851 were judged on construction and fine stitching. Bright colours were used to draw attention to the craftwork. These handmade boots and shoe, made for the 1851 exhibition, came from the London workshop of Robert Dixon Box. The shoe has its original wooden last in place.

Opposite and right: A strong interest in historical styles led to an enthusiasm for fancy dress during the 1800s. Queen Victoria's shoemakers, Gundry & Sons, made this Tudor style shoe for one of her children. The watercolour by Franz Haver Winterhalter was commissioned in the 1850s and depicts her son Prince Arthur dressed as Henry VIII.

Courtesy Hildegarde Fritz-Denneville Fine Arts, London

The late 1800s

Above: These silk court shoes were handmade in about 1885 by Henry Marshall, a London bespoke shoemaker. The curved construction of the new heel was a revival of styles from the previous century and was christened the louis.

Opposite: The 13 straps of this barrette boot are fastened with buttons and embroidered with jet beads. The multiple straps created a decorative effect and gave a tantalising glimpse of a stockinged leg. The boot is one of a pair made in the 1890s by the Joseph Box company in London.

A SEWING MACHINE for leather was in use by the 1850s and in the 1860s machines were developed for sewing on soles, for riveting, for making turnshoes and welt sewing. Much of this technology was invented in the USA. Machines were gradually developed to carry out each process, and by the end of the century most shoes were made in large factories. The personal relationship between shoemaker and wearer disappeared except at the most expensive end of the market.

Increased sporting activity saw manufacturers strive to make shoes that fitted well, while advances in technology and production led to a more diverse range of shoes and boots. Popular styles for men included ankle boots as well as laced shoes such as the oxford, the derby and the brogue. With the oxford style, the vamp was stitched to the quarters, which were joined by lacing. In the derby style, the vamp continued under the quarters to form a tongue over which the laces were tied. The brogue developed as a type of oxford, with perforations at the vamp and quarters. The perforated style of decoration was adapted from traditional Scottish footwear.

From the 1860s shoes became popular again for women and heels increased in height. Leather or cloth was used for the upper. The most popular style was a slip-on shoe known as a court or pump. A version of the court was the cromwell, loosely based on historical styles, which was cut high at the instep and decorated with a buckle. Other styles were laced or fastened over the instep with a strap and decorative effects were created by multiplying or crossing the straps. Embroidery, bows, buckles and beading were often applied at the toe and colours were varied, especially after the invention of bright synthetic dyes in the 1860s. More sensible walking boots and shoes in robust leather were also available, reflecting the increasingly active lives women were leading.

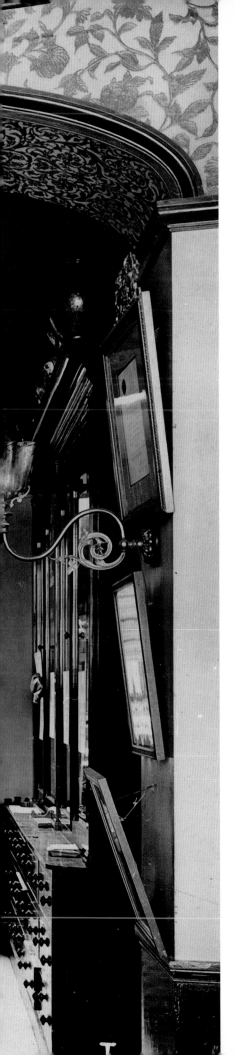

Left: Shops specialising in shoe retail appeared in major cities from the 1860s onwards. The photograph shows the interior of the Joseph Box shoe shop in Regent Street, London, in the late 1800s. A range of boots and shoes for both adults and children is displayed.

Courtesy City of Westminster Archive and Geremy Butler

Above: Boots for women became increasingly elaborate from the 1850s onwards, partly due to the introduction of machinery. This buttoned boot is one of a pair made for the Princess of Wales in the 1880s by Gundry & Sons, London. The quilted silk upper, which resembled a gaiter, was machine sewn.

Above: A developing interest in sport encouraged shoemakers to produce special footwear. This leather oxford-style shoe with a sole and heel made of rubber was suitable for tennis, a popular new sport in the late 1800s. It was made in 1886 by the Joseph Box company in London.

Opposite: Shoes with red heels were once symbolic of status and were favoured by Louis XIV (1643–1715) and his courtiers. This leather oxford is one of a pair made at the Joseph Box company in London about 1890.

The 20th century to now

*T*HE MAIN STYLISTIC trends in women's shoes in the 1900s include the strap shoe worn in the twenties, the platform sandal of the 1930s which developed from beachwear, the stiletto heel of the 1950s, the 'pop' styles worn in the 1960s, and the nostalgic revival of the platform-soled shoe in the 1970s. The increasing emphasis on casual wear has created a greater diversity of styles, and shoe design since the late 1970s has been characterised by a plurality of styles and a continual revival of past styles.

Today, bespoke or made-to-measure shoemakers mostly produce men's shoes in styles that had evolved by the end of the 1800s. However, the fashion industry has provided scope for individuals to demonstrate skills not only in the craft of shoemaking but in innovation in design and use of materials. Noteworthy shoe designers include André Perugia, Salvatore Ferragamo and Roger Vivier.

The ready-made shoe industry specialising in the fashion market was dominated by the USA in the early 1900s, but after World War II the industry became global. To compete in the world market, Italian shoemakers overhauled their traditional workshops and modernised. Nowadays leading shoe designers work from their studios in the fashion centres of Paris, London and New York, and some still have their designs made up in the factories of northern Italy where they maintain control of their product by close liaison with skilled shoemakers.

In the last decade the contemporary shoe market has been dominated by sports and 'lifestyle' shoes often manufactured outside these traditional regions. New plastics and finely tuned manufacturing techniques have fundamentally changed the nature of the shoe. From the 1980s a variety of styles catered for activities such as aerobics, jogging, basketball and skateboarding. The shape and detail of these styles have now become fashion statements, often linked to popular culture, music genres and specific subcultures. There are even websites with huge followings where the different styles are promoted and discussed. Select designs are launched in limited editions for truly serious sneaker collectors.

Opposite: Some of the most innovative shoes of the 20th century were created in the 1960s when shoe designers experimented with shapes and non-traditional materials. These metal sandals called 'Luna' were designed by Camille Unglik (b 1940) for the Swiss company Bally in 1970.

1900 to World War II

Above: In this fashion illustration from the magazine *Gazette du bon ton*, a flapper holds high a shoe by André Perugia, a *bottier* who worked with couturiers Paul Poiret and Elsa Schiaparelli. Originally from Nice, Perugia moved to Paris in 1921 where he made shoes for celebrities and gained recognition in the fashion press.
'Le Bel Ecrin', *Gazette du bon ton*, No 8, France, 1924–1925. Courtesy Bally Shoe Museum

Opposite: By 1910 shorter skirts revealed women's feet as never before, and shoes played a more obvious role in fashionable dress. The pre-war taste for luxurious shoes is exemplified in this beaded leather shoe of about 1910 made by the London department store Harrods. The silk stocking with butterfly motif was made about 1913.

THE RANGE OF READY-MADE shoes available for women at the beginning of the century reflected the more active lives women were leading, a tendency reinforced by the demand for women's labour in World War I. Lace-ups, strap shoes, courts or pumps and boots both laced and buttoned, were worn depending on the occasion. Of all the styles available, a shoe with a strap fastened with a single button and a waisted heel became the most popular style worn during the 1920s. The style was worn with the new short skirts and was practical for dancing. Made from a variety of brocaded silks and exotic leathers, and often with decorative details such as heels embellished with paste stones, these shoes reflect the taste for luxurious fashion which culminated in the art deco style of the International Arts Exposition in Paris in 1925.

Paris was the undisputed centre of fashion and those that could shopped at its boutiques and had shoes made up for them by *bottiers*, the bespoke shoemakers who worked in collaboration with the couturiers. The shoemaking workshop established in the mid 1800s by Jean-Louis François Pinet (1817–1897) maintained its reputation for fashionable shoes, but in the 1920s the most noteworthy shoemaker was André Perugia (1893–1977) who gained individual recognition for his bespoke shoes made with technical precision in unusual shapes and luxurious materials.

The influence of Paris was challenged by Hollywood in the inter-war period as the screen stars presented a type of glamour that people wanted to emulate. Styles of shoes associated with Hollywood suavity include the two-tone oxford for men and the low-cut high-heeled shoe for women. Influences also came from casual wear as the wooden platform sandal, worn first as beachwear, became popular in the 1930s, as did a slipper-like leather shoe for men which became known as the loafer or moccasin.

The ready-made shoe industry was dominated by American companies in the first half of the century. I Miller, Delman and Palter employed freelance designers and exported high-fashion shoes throughout the world. Italian shoemaker Salvatore Ferragamo (1898–1960) went to the USA in 1914 to study its ready-made system but ended up rejecting mechanisation to make bespoke shoes for Hollywood screen stars. He returned to Italy in 1927 and established a workshop in Florence which became famous for its craftsmanship and innovation; his most influential innovation was the wedge platform shoe which he made initially in cork as a response to restrictions on the use of luxurious materials imposed on Italy by the League of Nations in 1935. Antecedents of the platform shoe are found in the elevated platform sole shoe worn in Europe in the 1500s, called a chopine in England, and referred to in Shakespeare's Hamlet: 'Your ladyship is nearer to heaven than when I saw you last, by the altitude of a chopine'.

Above: Shoes made in the 1920s have a characteristic silhouette and typically featured luxurious materials such as snakeskin and brocaded silk. These shoes were purchased by a Sydney woman on a trip to Europe in 1925. The black and red shoe is by Pinet.

Left: Two-tone lace-up shoes known as spectator or co-respondent shoes were a craze in men's fashions in the inter-war period. The style was popularised by Hollywood and American jazz musicians as well as the Prince of Wales who wore them for playing golf. This shoe was made in Austria in the 1930s.

Opposite: Low-cut high-heeled shoes complemented the bias-cut evening dresses fashionable during the 1930s. It was a look popularised by the glamorous screen stars of Hollywood. These shoes have art deco styling and were designed by Palter de Liso in New York and made for the Sydney department store David Jones Pty Ltd.

Opposite: This mule, similar to an oriental *babouche* with its upturned toe, was patented by Italian shoemaker Salvatore Ferragamo (1898–1960) in the late 1930s. It features the wedge heel, an innovation in heel shape introduced into fashion by Ferragamo in 1936. The 'wedge' was one of the most popular shoe styles for women during the 1940s.

Above: Challenged by League of Nations sanctions imposed in 1935, Italian shoemaker Salvatore Ferragamo developed a platform heel and sole made of cork. With these sandals of 1938, Ferragamo padded the leather straps and covered the platform with hand-painted silk embroidered with tiny glass beads.

World War II and the 1950s

LUXURY INDUSTRIES WERE curbed during World War II and stylistic changes in fashion came to a standstill. Styles introduced in the late 1930s, in particular the platform court shoe, remained fashionable throughout the 1940s. Innovation was limited to the use of materials not essential for the war effort: for example wood, hinged for flexibility, was used for making the soles of shoes.

The revival of Paris couture after the war was led by Christian Dior who in 1953 began a collaboration with French shoe designer Roger Vivier (1913–98). Vivier designed bespoke shoes for Dior's couture collection, followed by a mass-produced collection with Vivier's name on the label with Dior's. Vivier's shoes were noted by the fashion press for their seasonal changes in toe and heel shapes, and his designs, such as the changing silhouettes seen in Dior's couture collections, received attention. Along with other French and Italian designers, Vivier developed an increasingly low-cut court shoe with a narrow heel which by 1952 was known as the stiletto. Fashionable dress for women in the 1950s required careful attention to one's choice of accessories. To be stylish it was essential to wear a tailored suit or dress with matching gloves, hat, handbag, umbrella and stiletto-heeled shoes to complete the look.

Most men continued to wear the styles which had developed by the end of the 1800s, but distinctive shoes worn by sub-cultural groups created fads in shoe styles. 'Brothel creepers' were shoes with platform crepe soles and were part of the Teddy Boy dress in the late 1940s, and winkle-pickers were first worn in the late 1950s by Mods. The distinguishing feature of the winkle-picker was its elongated sharp toe, reminiscent of a Medieval shoe fashion known as the poulaine.

In the post-war period the shoe industry was dominated by large companies such as Charles Jourdan in France, H & M Rayne in England, Delman and I Miller in the USA, and Bally in Switzerland. To compete in a global market, artisan shoemaking workshops in Italy overhauled traditional work practices and mechanised their workshops. The 'Made in Italy' promotional movement and the modernisation of the Italian shoe industry in the 1950s did much to improve the economy of northern Italy.

Above: Advertisement created by René Gruau in 1960 to publicise shoes by Roger Vivier for Christian Dior.
Courtesy Christian Dior

Opposite above: Along with other French and Italian designers, Frenchman Roger Vivier developed an increasingly low-cut shoe with a high and narrow needle heel which by 1952 was known as the stiletto. This shoe by Vivier dates from 1961.

Opposite below: The two-tone court shoe was created by Parisian *bottier* Raymond Massaro (b 1930) in 1957 for the couturier Coco Chanel who wanted a more comfortable alternative to the fashionable stiletto. The shoe's beige body and heel optically extend the line of the leg while its black toe cap makes the foot appear smaller. The look has appeared in every Chanel collection since; this slingback by Massaro dates from the mid 1960s.

Opposite: The slender heel of the 1950s stiletto was usually complemented by a pointed toe which was often jewelled for evening wear. These stilettos dating from the late 1950s have been embellished with beads, pearls and diamantes. From bottom to top, they were made by Vivier for Dior in France, Ferragamo in Italy and Rayne in England.

Above: These shoes were worn by an Australian diplomat; they exemplify the conservative nature of men's fashions since the 1900s. With the exception of the ankle boot (left), these shoes were made to measure in the 1960s and 1970s by McAfee in London. The shoe on the far right was worn with formal attire.

Pop and nostalgia, the 1960s and 1970s

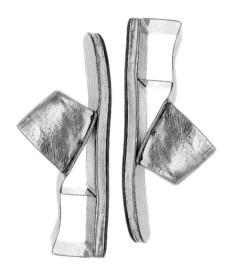

Above: Some of the most experimental fashions created in the 1960s and 1970s used non-traditional materials such as plastic and metal. These metal mules were designed by Camille Unglik for the Swiss company Bally in the early 1970s.

Opposite: When the mini-skirt became fashionable in the mid 1960s, boots leapt to popularity once again. The innovation of combining a shoe and stocking to form a 'stocking boot' was developed by New York shoe designer Beth Levine (1914–2006) in 1967. This version has a clear acrylic heel attached to a plastic shoe overlaid with a stocking worn up to the thighs.

IN THE 1960s fashion was primarily influenced by the post-war 'baby boomers' who rejected the values of their parents. The formal style of dressing dictated by Paris couture was abandoned and, with the young in mind, designers created new clothes such as the mini-skirt, hot pants, trousers and boots which were made from the latest, often synthetic, materials, in dazzling and vibrant patterns and colours inspired by pop art. For the first time since the 1700s, men wore clothes that were flamboyant and colourful. These clothes suited the more informal lifestyle that young people wanted to lead and were a means of expression. Confidence in the future, developments in space exploration and advances in technology inspired fashion design.

Low-heeled shoes with square toes replaced the stiletto for women, and shoes made from brushed pigskin and corduroy for men became popular, reflecting the move to casual dressing. With the introduction of the mini in the mid 1960s women's legs were more exposed, and fashion designers such as Mary Quant (b 1934) in London and André Courrèges (b 1923) in Paris created ankle- and knee-high boots to accentuate the new look. Quant's boots were made of brightly coloured plastic using an injection-moulded method developed in Britain in the mid 1950s, while Courrèges used white patent leather to complement his 'space age' look. Some of the most avant-garde clothing was created by Paco Rabanne (b 1934) who used alternative construction techniques in his garments, including shoes, made from plastic or metal.

By the late 1960s disillusionment with contemporary life and anxiety about the future led young people to embrace the hippy culture with its anti-fashion philosophy, disregard for the new, and fondness for ethnic dress. Colourful footwear was worn with long hair and denim jeans. The revival of the platform sole was an integral part of the nostalgia of the period. It was a rage in shoe design that continued up to the mid 1970s and platform soles were enthusiastically worn by both men and women.

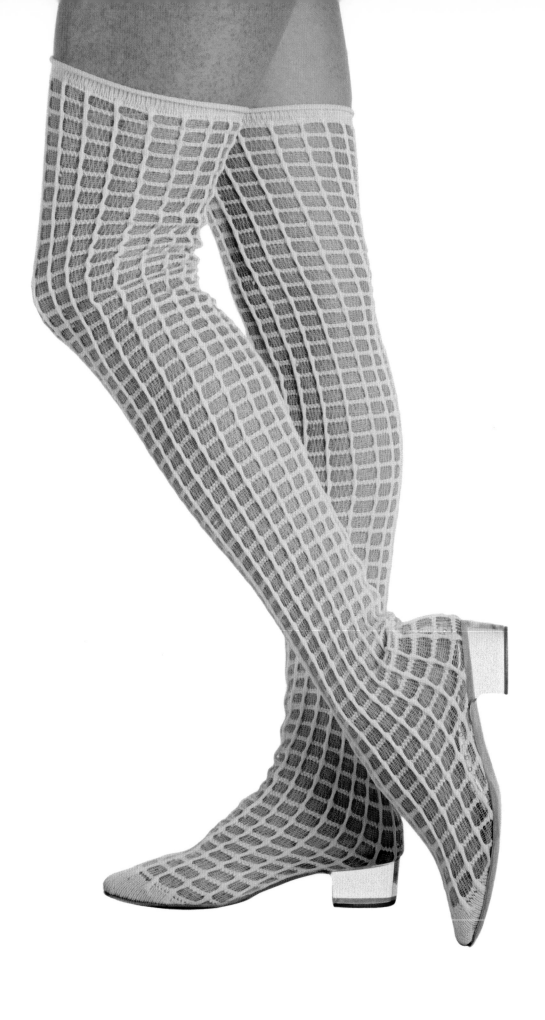

Above: The freeing up of dress codes in the 1960s allowed for more innovative shoe designs. Some of the best were by New Yorker Beth Levine who named these 1964 shoes 'Kabuki'. The lacquered wooden heel is inspired by Asian design but the shoes have a unique aerodynamic quality.

Left: These boots of 1965 by Mary Quant are made from injection-moulded plastic, a technique first used in Britain in 1956. They exemplify 'pop' fashion which epitomised a celebration of youth, new technologies and original clothing. The heels are stamped with a daisy, the Quant trademark.

Above: The hippy look was characterised by colourful nostalgic clothing. Ironically, the anti-fashion fondness for textiles and garments from the past was taken up by fashion designers. These boots made of patchwork silk reminiscent of a traditional quilt were designed by Beth Levine in the late 1960s for the upper end of the market.

Right: Platform sole shoes, a revival of the 1940s style, but taken to greater heights and worn by both men and women, were very popular during the early 1970s. They were an integral part of the nostalgia that characterised the hippy look. London designer Terry de Havilland (b 1938) created this pair of platform wooden clogs in about 1974.

Contemporary style

Above: The stiletto has been in and out of fashion since it was first worn in the 1950s. The design of Christian Louboutin's Pensée (Pansy) shoes of 1994 were inspired by the pop art of Andy Warhol and have a stiletto heel with a termination shaped like a paw. Louboutin (b 1964) calls them 'follow me shoes' and describes his shoes as pedestals for women.

Right: One of the most inventive shoe designers in the 1980s was Tokio Kumagai (1948–1987), whose quirky shoes masquerading as cartoon mice or racing cars continued a tradition of illusionary techniques seen earlier in the century, in particular by Elsa Schiaparelli in her surrealist inspired designs of the late 1930s. The uppers of 'Mouse' designed in 1985 are made of calfskin with leather trimming.

SINCE THE BREAKING of fashion's traditional rules and hierarchy, fashionable dress has been characterised by a plurality of styles. The value and meaning of clothing had changed in the 1960s, and culminated in the emergence of punk culture in 1976 with its total rejection of contemporary clothing values. In the last three decades, individual designers have worked on their own themes or philosophy of dress, taking inspiration from the street, from other cultures and the work of known designers, and particularly from the past. The recycling of fashions in shoes occurs again and again.

Despite the confusing diversity of styles some trends are evident. Simple flat-heeled shoes in dark colours have been popular with women since the early 1980s, reflecting the influence of casual wear and women's changing needs. Flat-heeled shoes based on footwear worn by Chinese and Japanese peasants were also teamed with new styles of dress introduced in the 1980s by influential Japanese designers based in Paris. Another major trend is the wearing of heavy looking boots and shoes inspired by traditional workwear. The popularity of 'Doc Martens' and Blundstone boots, particularly in the 1980s and '90s, are good examples of this trend.

The stiletto, the sandal, the court, the platform, the loafer and lace-up styles continue to be produced. No matter how impractical, many designers continue to create high-heeled shoes. With variations in materials and colours as well as changes in cut and detail, familiar styles can capture the mood of the moment. The recycling of fashions in shoes does not necessarily mean a lack of ideas, as the most inventive of contemporary shoe design comes from those designers who have given their creations a fresh interpretation.

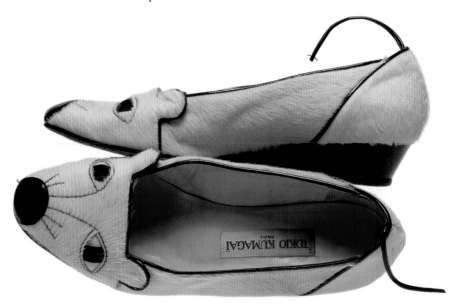

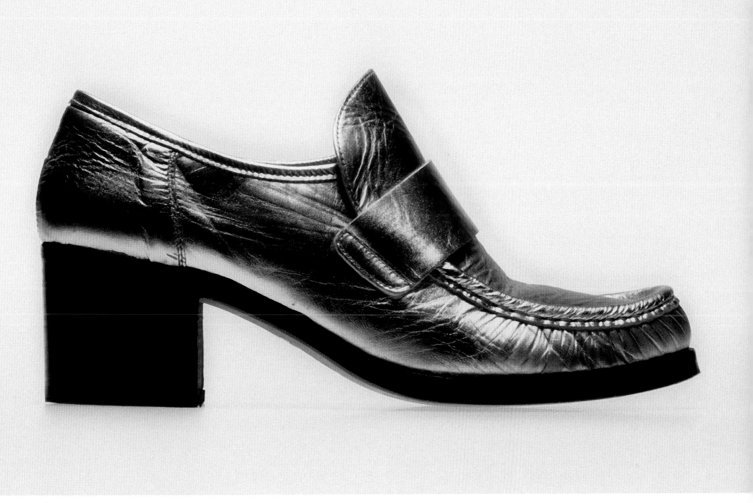

Above: The style known as a loafer has been popular since the 1930s when it first became fashionable. The 'Wannabe' loafer created by London based designer Patrick Cox (b 1963) in 1993, has exaggerated proportions and its egalitarian no-brand look parodies the loafer's status symbol image.

Right: Since the late 1980s shoe designers have been influenced by traditional workwear. Belgian fashion designer Dirk Bikkembergs (b 1962) collaborated with Flemish shoemakers who made work boots to create heavy-looking fashionable footwear with macho appeal. With these boots of 1996, eyelets and laces have been replaced with a stretch cord which is worn wound around the ankle and through a voided metal heel.

Above: Popular culture and streetwear are two of the many influences on high fashion. The craze for wearing sports shoes influenced the design of these shoes of 1996 by Parisian designer Christian Lacroix (b 1951). The uppers are made of reflective nylon and the sole and heel are of rubber.

Opposite: High-heeled shoes with platform soles are not new in fashion but British designer Vivienne Westwood (b 1941) took the platform sole to extremes, juxtaposing the traditional with the taboo in this design. Called 'super elevated gillies', these shoes were named after the traditional shoes worn by Scottish highland dancers, but the design owes more to the influence of fetish footwear.

Opposite: New methods of construction and materials for sports shoes have transformed the shoe industry. With injection moulding and bonding, many manufacturers have departed from traditional stitching techniques. Nike, established in the early 1960s, has been at the forefront of sports shoe research to enhance comfort, fit, agility and speed. These Nike shoes were made in the late 1990s.

Above: Australians Tull Price (b 1977) and Rodney Adler (b 1977) launched Royal Elastics in 1996 to do away with shoelaces. First launched in London, its style is a fusion of fashion and sport. The company was one of the first to manufacture limited edition collectable shoes. These Splice Clogs for women are from the 1999 range.

Above: Marc Newson (b 1963) was inspired by Russian cosmonaut boots when he designed the Zvezdochka shoe for Nike in 2004. The sock-like insert can be removed and the perforated upper allows the foot to breathe. A limited edition, the shoe was launched in New York. Australian designer Newson believes the aeronautical industry drives technological development in design.

Opposite: The use of sturdy denim for these glamorous stiletto boots, with woven designer logo, typifies Galliano's 2000 theme of 'turning couture inside out'. John Galliano (b 1960) graduated from Central St Martin's School of Art in London to become head designer for Givenchy in 1995 and Christian Dior in 1996.

The Australian shoe industry

A SHORTAGE OF SHOES was a constant problem for early European settlers in the colony. Bare feet were not officially acceptable although indigenous Australians recognised that there was little need for shoes in this mild climate. In *Fashioned from penury* (1994), Margaret Maynard noted that advice to settlers, letters home and the much-repaired state of surviving footwear all confirm the severe shortage, especially of women's shoes. Colonists initially depended upon imports. In 1793 the vessel *Speedy* delivered 6400 pairs of shoes to Sydney. But supplies did not always arrive: merchant ships were sometimes lost, or shoes badly mildewed in transit. The ship *Sydney Cove,* laden with bundles of men's goatskin shoes from India, sank in Bass Strait off Tasmania in 1797.

By the 1820s, when there were more women immigrants in the colony and survival was less of a preoccupation, social gatherings became increasingly fashionable. Of the preparations for the King's birthday ball at Government House in 1826, the *Sydney Gazette* noted: 'Tailors and tailoresses, shoemakers and shopmen were in great requisition up to the auspicious moment'.

Imports were sporadic and the merchant shipping trade was often speculative. When a ship arrived, locals on the wharves bartered wheat, hides or other produce for manufactured goods such as men's shoes and clothing. Shoemakers soon set up business in Sydney.

As the gold rush gained momentum, imports of all kinds increased tenfold and Australia became a major export market for British shoe companies. One of those, C & J Clark, still manufactures in Australia. The great exhibitions held in London in the 1850s and 1860s were planned as marketing opportunities for Britain, but they proved even more beneficial to Australia, creating a lucrative trade in fine goods and raw materials, and promoting Australia as an attractive country for industrious immigrants and investors.

By 1900 shoe manufacture had become a significant industry in Australia and the industry was almost self-sufficient in materials. A network of supporting industries evolved, especially in Victoria where protective tariffs were imposed even on goods imported from the other colonies. Shoe manufacturing was a major employer of men and women, and many (but not all) manufacturers produced quality goods. By 1959, a massive 35 million pairs of boots and shoes were being manufactured each year in Australia.

Large-scale manufacture diminished dramatically when federal tariffs were progressively reduced from 1973. The industry at present is struggling but companies with small niche markets are surviving.

Opposite: Australian shoe manufacture today includes a number of young designers who are inspired by the traditional craft of shoemaking. Using crocodile skin harvested in Queensland, Sydney shoemaker Donna-May Bolinger (b 1960) hand dyed the leather and made these men's oxfords entirely by hand.

Early colonial times

EARLY IN THE 1800S shoemakers began to establish themselves in Sydney. In 1809 a Mr Hayes advertised in the *Sydney Gazette* for a 'Person to superintend a Boot and Shoe Factory'. His factory offered 'hessian boot legs, shoes and boots, boot top leather, ladies' and children's shoes, colours, russet calf skins for ladies' shoes …'.

With limited supplies, quotas of clothing were issued by the government – for convicts one pair of shoes every six months, for woodcutters or bullock drivers every three months. When imports did not arrive, shoes were made locally, but in 1831 Governor Darling ordered 10 000 pairs of shoes for the following year from Britain, arguing that this would be cheaper than manufacturing them in Sydney where production was less efficient.

James Thearle, a third-generation shoemaker, emigrated from England in 1838. He worked as foreman with James Vickery of George Street, Sydney, who had a government contract to make boots for convicts. Several examples of Thearle's work were donated to the museum by his great grandson, Frederick Thearle, in 1958.

Shoemaking and mending were easily transportable skills and it was a popular occupation amongst new immigrants. It was traditionally a man's profession since the strong linen threads were very hard on the hands. However a whole family could contribute, children running errands, women and older girls sewing the uppers, men and boys attaching the soles. Skills were often passed down through several generations. Shoemakers were often employed on large country properties in the 1820s. By 1828 shoemaking was, after carpentry, the largest occupation, with one shoemaker per 236 inhabitants.

Above: Local shoemakers promoted their businesses with advertisements like this from an 1838 advertising catalogue.

Courtesy Joseph Lebovic Gallery

Right: The single miniature kangaroo-hide wellington boot, made in Sydney by James Thearle, boasted '42 stitches to the inch' (17 stitches per cm). It was displayed in Abbey's shoe shop window for 12 months and a reward of £50 was offered to anyone who could make a boot to complete the pair. Also pictured are a miniature last and an unfinished boot from the Thearle collection.

Above: The small clog overshoes (right) were said to have been worn by the first girl immigrant to step ashore at Moreton Bay, near Brisbane. They were tied on over fashionable shoes as protection. The solid leather clogs were for day-to-day wear, with brass nails and horseshoe-shaped irons nailed to the soles. It is difficult to distinguish the place of manufacture of early shoes in the collection as the British and Australian makers used the same techniques.

Right: The colony soon began to develop profitable exports of local materials. Kangaroo hide was found to be stronger for its weight, finer and more comfortable for shoes than other leathers. It is still highly prized today by bespoke shoemakers. Wattle bark proved an excellent leather tanning agent and was an export earner until the technique for chrome tanning was perfected in 1900.

Australian Leather Journal, 15 June 1903. Courtesy State Library of NSW

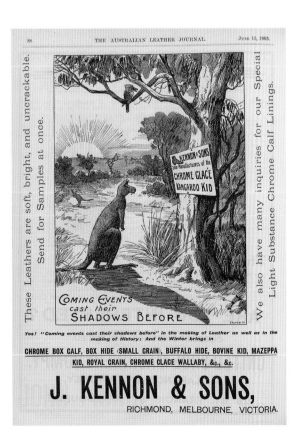

Establishing a footwear industry

Above: The elastic-sided boot first designed by R M Williams in 1932 is exported worldwide. Constructed with only one back seam, the qualities of water resistance, comfort and toughness qualified it to become a household name among rural communities. The style epitomises the carefree macho Australian male, but these days 'RMs' can also be seen on international catwalks.

IMMIGRANTS WITH NEW skills played a major role in shaping the industry, as did improved techniques and new materials. In the 1860s new machinery from overseas changed the nature of shoemaking. The 1880s saw machines sewing 900 stitches a minute and the evolution of mechanised mass production. After the 1886 Colonial and Indian Exhibition the *Art Journal,* London, reported of Australia: 'The Grand United Oceania of the future, the last born child of the old country, is doubtless destined to become one of the greatest manufacturing centres of the world'.

It was America, however, that led the way with improved techniques, machinery and design for fashionable shoes. Liberated American women stepped out in assertive black and brown boots. With the help of mail order catalogues, imports gained momentum in Australia. Australian manufacturers and workers fought fiercely to protect their local industries, a key issue leading up to Federation in 1901. The federal tariff legislation of 1902 put an end to American and British imports; shoe imports shrank to almost nothing and Australia's shoe industry survived.

Above: Retailers were to have a powerful influence upon styling and local manufacture when the volume of imports increased after World War II. Edward Meller retailed high-fashion imported shoes in Sussex Street, Sydney, but also made to order for clients, including this pair of platform shoes made for Lady Hurley in the late 1930s.

Right and opposite: Leisure wear became more fashionable in the 1920s. In a complete departure from traditional materials, Barnet Glass, a Melbourne company, began experimenting with byproducts from the gas industry in the 1880s to manufacture waterproof clothing. When it later merged with Dunlop, the company's range included these moulded rubber surf shoes, as well as sport shoes, plimsolls, galoshes, and wellington boots. Volley shoes, its most famous brand, were strictly for tennis – the era of sport shoes for general leisure wear had not yet arrived.
Home magazine, 2 November 1931.

The prosperous years of manufacturing

Above and opposite: The availability of powerful new cements made a fine pointed toe possible. This was difficult to achieve with a welt construction. Raoul Merton manufactured superb men's high fashion winkle-pickers. 'Of comfort you're certain when you're wearing Raoul Merton' was the slogan when Australian TV personality Graham Kennedy offered the shoes as prizes on his 1960s quiz show; they were the first shoe company in Australia to use radio and television marketing.

Right: Shoes from the 1930s. Selby Shoes (left), set up as a manufacturing company by David Jones in 1934, supplied all their stores around Australia until 1977. Like Westbrook and Mason (centre), their shoes were known for their fine quality. The Public Benefit Bootery (right), operating from 1908 to 1996, initially sold shoes at the one price of 10 shillings. As the name suggests, they offered good quality 'within a working girl's budget'.

THE RETAILING OF SHOES and the marketing of brand names became more sophisticated after 1900. Shoes with the company's ring stamp on the sole, or the name on the inner sole lining, often absent on earlier Australian shoes, inspired product loyalty, particularly for expensive, stylish shoes made for department stores. Other stamps promoted a technique or quality materials: the Goodyear welt, or the Dunlop heel or the Selby Arch Preserver. Marketing aimed to create and capture a shoe-buying elite. An advertisement from Parker Shoes in a 1934 copy of the *Home:* 'Your personality ... what an intangible, lovely thing when nothing disturbs it. But what a serious matter when it is threatened! You can save it by refusing to countenance commonplace shoes'.

Retailers were to have a powerful effect on styling and local manufacture when the volume of imports increased after World War II. Joe Goldberg's Melbourne company, Voma, was a giant of the import and retail business in the 1960s and 1970s and continued until 1996. He had the Ferragamo agency and advertised his locally manufactured product as 'styled in the US'.

Opposite: Merivale Hemmes launched the House of Merivale in Castlereagh Street, Sydney, in the 1960s. In a new approach to retailing, she created a total look from head to foot for the 'young generation' who flocked to the shop. Merivale had three factories in Sydney producing shoes from her designs. Platform soles returned briefly to fashion during the 1970s and were worn by both men and women. They have made occasional reappearances in women's fashion since then.

Above: Thongs had a dramatic impact on the Australian footwear trade in the late 1950s, when thousands were imported from Asia. They offered welcome relief from the painful stilettos and winkle-pickers of the time. Australian feet are wide, needing an additional 1/8 inch (almost 3 mm) width on the American Brannock fitting system. It is thought that a lifestyle of bare feet and open shoes in summer has caused the Australian foot to become wider.

The industry changes

Above: Computer technology is used in the Australian shoe industry, predominantly for design, grading and cutting purposes for mass production. The CAD/CAM design system allows a designer to picture a new style and also gives a complete specification of materials used in the construction of the shoe upper.
Courtesy J Robins & Sons Pty Ltd

Opposite above: This handmade shoe, called 'Prince of autumn leaves', was designed in 1994 by the Pendragon Boot Company, Brisbane. The fantasy design has since been reinvented as sandals and high-heeled shoes and featured in the Warner Bros film *Peter Pan* in 2002.

Opposite below: Film and theatre provide interesting work for a limited number of shoemakers. Jodie E Morrison (b 1953) has been making shoes in Sydney for the performing arts for over thirty years. During this time she has tackled shoes for a wide range of historical periods and design themes. This shoe was made for the production *Phantom of the Opera*, first staged in Melbourne in 1991.

AFTER STRONG GROWTH through the 1950s and 1960s, protective tariffs on clothing and footwear were reduced by 25 percent in 1973. This forced manufacturing companies to radically reassess production. J Robins & Sons, founded in 1873, successfully changed from large-scale mass production to twelve-person 'just-in-time' systems. With this method, more styles could be produced concurrently and rejects were eliminated. The company now produces 35 to 40 percent of Australian mainstream fashion footwear and its production system rivals any in the world. They also manufacture for niche design studios and are able to minimise production time.

Since the 1980s global sport shoe companies have taken the lion's share of the Australian shoe-buying dollar. They manufacture in third world countries at costs and under conditions with which Australia cannot compete. In April 1997 an agreement was reached in the USA to create a 'no sweat' swing tag to identify companies which adhere to a set of minimum wages and conditions in manufacturing.

From its inception in 1906 the School of Footwear in Ultimo, Sydney, has trained many of the best shoemakers in the trade. The late Bill Delaney, head teacher until 2000, recorded the closure of many businesses since tariff reduction commenced, in his work 'From Erskineville to Ultimo: a history of the NSW School of Footwear' (1996).

He concluded 'The face of the industry has changed remarkably. It is indeed a great deal smaller in size with the majority of factories employing fewer than 30 people. However the footwear produced is of a very high standard with many factories exporting'.

Australians are the fourth biggest buyers of footwear in the world, buying 3.5 pairs of shoes a year per head of population. Sadly, only a few of these are now made in Australia.

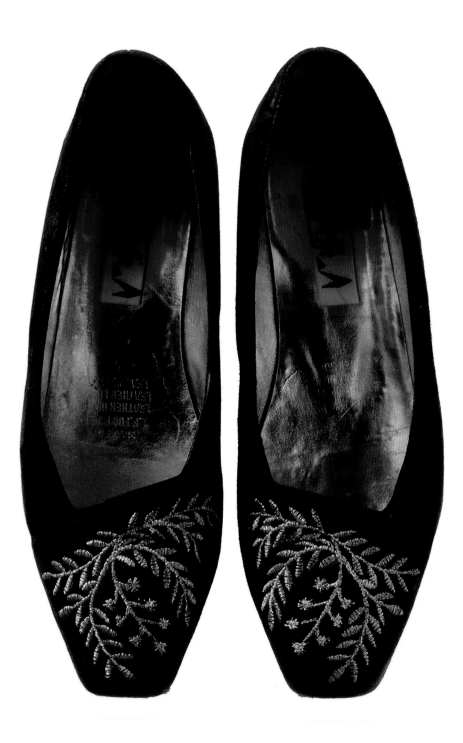

Left: For many years Leonie Furber has been influential in Australian shoe design. Working as a stylist (not a maker) and assisted by the French designer Philippe Modèle, she launched the first range of her Elle Effe label in 1989. A popular style from that range featured the wattle flower in gold thread on black suede, embroidered by the Sydney company Princess Embroidery.

Opposite: Striking effects with screen printed leather have become a unique trademark of Donna-May Bolinger's designs. Her 'bench-made' shoes are beautifully crafted. These shoes feature an unusual last shape – the key to innovative shoe design.

Shoe listing

The Joseph Box shoes, Museum numbered H4448, were all purchased in 1942.

All measurements are given as heel height x shoe length and in the case of boots, leg length including the heel.

TIAS refers to Tax Incentive for the Arts Scheme.

Front cover: **Woman's barette boots, Joseph Box, England, about 1896**
patent & glacé kid; leather; linen lining; jet beads
pointed toe; 13 straps buttoning at front; louis heel
woven: By appointment Joseph Box/ [the Royal Arms]/187 Regent St London
70 x 225 x 328 mm high
H4448-108

Page 2: **Woman's button boots, prize work, maker unknown, England 1870–1875**
morocco, patent and kid leathers
58 x 235 x 210 mm high
H4448-505

Page 5: **Woman's shoes 'Kabuki', Beth Levine, NY, about 1964**
leather; satin; lacquered wood
pointed toe; platform sole; sculptured wedge heel
label: Saks/Fifth venue/Fenton
18 x 255 mm
97/185/2

Foreword

Page 8: **Woman's slip-on shoes, John Thomas for Henry Marshall, England, about 1885**
silk; leather; kid & linen lining straight; turned construction; narrow oval toe; covered louis heel
paper label printed: By appt to her Majesty [Royal Arms]/ Henry Marshall, late Pattison, Bootmaker HRH Duke of Edinburgh, HRH Princess Louis of Hesse, HRH Princess Christian, 154 Oxford St London
79 x 243 mm
H4448-79

Traditional shoes from around the world

Page 18: **Man or woman's *waraji* or sandal, Japan, 1850–1900**
plant fibre
straight; interwoven; centre front and side loops with tie strings
15 x 215 mm
H4448-1043

Woman's *geta* or platform shoe, Japan, 1850–1900
velvet; lacquered wood; straw inverted-V straight sandal; possibly cemented construction; round toe; finely woven reeds (*zori*) on wood-block platform with two stilts underneath tag on sole: '3[8]3
54 x 216 mm
H4448-1041

Page 19: **Boy's slip-on cloth shoe, China, 1850–1895**
embroidered and appliquéd silk damask; cotton; leather
straight; forward jutting prow over rounded toe, platform
60 x 170 mm
H4448-1004

Man's slip-on basket-woven shoe or mule, China, 1850–1895
split rattan; soft plant fibre (?rushes)
straight; plaited and woven construction; square rounded toe; no heel
15 x 280 mm
H4448-1020

Page 20: **Woman's slip-on 'lotus' shoe for a bound foot, China, 1880–1895**
embroidered silk brocade; metal thread; cotton lining; leather; wood
straight; upcurved needlepoint toe; covered wooden heel and platform
70 x 140 mm
H4448-1005

Man's toe-ring sandal, southern India, mid 1800s
leather; copper
thonged construction; footshape toe, rounded stacked heel
30 x 240 mm
H4448-71

Page 21: **Man's toe peg sandal, India, mid to late 1800s**
ivory; metal; bronze
straight; one-piece platform with stilts; separate metal toe peg; surface engraving
22 x 248 mm
H4448-1015

Man's mule, India, late 1800s
leather; shagreen; metallic thread; beetle wings
straight; turnshoe construction; pointed toe curled over, stacked heel
24 x 254 mm
H4448-1012

Page 22: **Woman's mule overshoe, Turkey, mid 1800s**
leather; leather lining; felt sock; metallic thread
straight; possible rand construction; machine stitched; pointed upturned toe; high toespring; no heel
40 x 250 mm
H4448-1010

Woman's bath clog, Palestine, c 1890
wood; mother of pearl and silver inlay
straight; one piece carved construction; square toe; nail holes for missing toe-band; flaring stilt heel
60 x 230 mm
H4448-1037

Page 24: **Man's ankle boot, Palestine, late 1800s**
leather; rawhide; metal; strap and buckle fastening
straight; turnshoe construction; pointed toe curled over and nailed to vamp; no heel
0 x 315 mm
H4448-1030

***Yezmeh* (calf-length riding boots), Damascus, Syria, 1895–1905**
cross-boarded goatskin; camel; rope; metal; cotton; leather lined
straight; tunnel stitching through sole; blunt pointed upcurved toe; iron heel
15 x 282 x 511 mm high
H4448-1035

Page 25: **Man's sandal, Palestine, 1895–1905**
rawhide
thonged construction; inverted V-shape; knotted toe thong; footshape toe; no heel
0 x 252 mm
H4448-1033

Hausa man's toe-thong sandal, Africa, mid to late 1800s
leather; rawhide
straight; thonged construction; pointed toe, no heel
0 x 250 mm
H4448-1017

Page 26: **Cree Indian moccasin, Canada, mid to late 1800s**
leather; beaded felt; velvet
straight; gathered and stitched construction; oval toe; oval seat, no heel
0 x 270 mm
H4448-1003

Child's slip-on shoe, Greenland or Iceland, 1830–60
sealskin; fur lining
straight; turnshoe construction; square toe; no heel
0 x 146 mm
H4448-1029

Page 27: **Woman's mule clog, North Italy or Portugal, late 1800s**
wood; leather; corduroy; sheepskin; leatherette lining
straight; nailed construction; pointed toe; heel carved as part of sole
57 x 272 mm
H4448-279

Infant's or miniature clog, France, late 1800s
wood
carved in one piece; pointed with crescentic hook; heel carved as part of clog
10 x 127 mm
H4448-1045

Fashionable footwear: the 18th and 19th centuries

Page 28: **Woman's miniature mule (prize work), England, about 1851**
silk brocade; metallic fringe; leather
straight; turned construction; needlepoint toe; louis heel
70 x 159 mm
H4448-531

Page 30: **Girl's tie shoe, England, 1700–1710**
leather; wood
single straight; rand construction; wedge louis heel
31 x 160 mm
H4448-1

Page 31: from left
Woman's buckle shoes, 1720–1730
silk brocade; leather; wood; kid lining
straight; rand construction; upturned needlepoint toe; covered louis heel
61 x 249 mm
H4448-54

Woman's tie shoes, England, about 1710
linen (embroidered in early 1600s and reused); leather; wood; linen & leather lining
straight; rand construction; upcurved, blunt pointed toe; covered louis heel
58 x 239 mm
H4448-7

Woman's tie shoes, England, 1700–1710
silk brocade; leather; wood & morocco; kid & linen lining
straight; rand construction; upcurved needlepoint toe; covered louis heel
75 x 239 mm
H4448-52

Woman's tie shoes, probably England, about 1700
embroidered linen; leather; wood; kid lining
straight; rand construction; covered louis heel
60 x 220 mm
H4448-55

Page 32: **Woman's mule, probably France, 1775–1780**
silk brocade; leather; wood; leather lining
single straight; rand construction; upcurved oval toe; right-angled louis heel
60 x 238 mm
H4448-50

Page 33: **Woman's slip-on shoes, England or France, 1787**
morocco; leather; wood; looped twill lining
turned construction; pointed upcurved toe; covered louis heel
50 x 264 mm
H4448–60

Page 34: **Man's buckle shoe, England, 1761**
kid & morocco; gold lace rose; leather; wood
single straight; rand construction; prow toe; covered heel
35 x 270 mm
H4448–10

Page 34: **Shoe buckles, England, 1780–1786**
silver; enamel; paste
51 x 90 mm
Joseph Box collection
H5736–22

Page 35: **Postilion boot, probably French, about 1790**
leather
single; thigh length; square domed toe; stacked heel
46 x 300 x 618 mm high
H4448–33

Page 36: from left
Woman's tie shoes, England, 1790–1820
leather; (ribbon not original); linen lining
straight; welted construction; needlepoint toe; stacked heel
7 x 231 mm
H4448–68

Woman's slip-on shoes, England, 1790–1810
morocco leather; wood; linen lining
straight; turned construction; needlepoint toe; wedge louis heel
24 x 234 mm
H4448–63

Woman's slip-on shoe, England, 1790–1795
kid & silk; leather; wood; linen lining
single straight; turned construction; needlepoint toe; wedge louis heel
36 x 242 mm
H4448–62

Woman's slip-on shoes, about 1795
printed kid leather
turned construction; needlepoint toe; wedge louis heel
25 x 255 mm
gift of K Agnew 1961
H6871

Page 37: **Woman's front-laced ankle boots, England, about 1804**
silk; leather; linen lining; cord lacing
straight; turned construction; stacked heel
18 x 248 x 200 mm high
H4448–11

Page 38: **Man's tie shoes, England, 1835–1840**
leather; grosgrain ribbon; leather lining
welted construction; upcurved square toe; stacked heel
34 x 262 mm
H4448–65

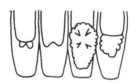

Page 39: from bottom
Woman's slip-on shoes, Gundry & Sons, London, about 1832
silk satin; leather; linen & kid lining
straight; turned construction; square toe
paper label printed: 1 Soho Square/ [Royal Arms]/Gundry And Son/ Boot & Shoemakers/to/Her Majesty/ and Their Royal Highnesses/Duchess and/Princess Victoria of Kent/and Her Serene Highness the Princess/ Hohenlohe Langenburg/London
0 x 243 mm
H4448–24

Woman's slip-on shoes, London, 1830–1850
silk; cotton tapestry weave (kilim); leather; leather lining
straight; turned construction; square toe
leather label printed: Fraser/ Manufacturer/Wholesale & Retail/172/ Ratcliff H'way/Corner of [...]/London
0 x 235 mm
donated by J I Horne
A6119

Woman's slip-on shoes, France, about 1830
silk satin; leather; kid & linen lining
straight; turnshoe construction; square toe
paper label printed: Thomas Moore Importer of Foreign Boots and Shoes. Wilton House Knightsbridge; stamp on sole: Mayer, Paris
0 x 265 mm
acquired in 1978
A7020

Woman's slip-on shoes, probably Austria, 1830–1860
plaited straw; leather; silk lining
straight; turned construction; square toe
0 x 255 mm
gift of Jacqueline Lillie 1986
86/491

Page 40: **Man's napoleon knee boots (prize work), Pattison, England, 1850–1870**
leather; patent; morocco; leather lining
welted construction; square rounded toe; stacked heel
notes: made by Pattison, Oxford St, London
10 x 255 x 370 mm high
H4448–515

Page 42: **Man's front-laced balmoral boot (prize work), William Walsh, England, about 1851**
patent & glacé kid; leather; silk lining
single straight; welted construction; square rounded toe; stacked heel
inscribed: William Walsh, maker, London 1855
notes: this boot won the prize medal at the Paris Exhibition 1855
27 x 245 mm
H4448–524

Page 43: **Woman's spring-sided boots, Joseph Sparkes Hall, London, 1837**
linen; metal springs covered with cotton fabric; leather
straight; turned construction; square toe; low heel
label inscribed: ... Sparkes Hall ... London
provenance: Jean Downes; Miss H D Butcher; Mrs William Dean Butcher, wife of Windsor doctor; Mrs [Lidded] lady in waiting to Princess Christian
3 x 135 mm
purchased 1994
94/88/1

Page 45: **Boy's fancy-dress rose shoe, Gundry & Sons, London, about 1846**
silk; mock Tudor cut-outs on vamp; leather; linen lining
single; turnshoe construction; wide toe
0 x 158 mm
H4448–27

Top, from left to right
Woman's side-laced balmoral boot (prize work), William Walsh for Robert Dixon Box, London, about 1851
patent leather; gold fringe; leather; linen, linen & silk lining
single right straight; rand construction; stacked heel
20 x 232 mm
H4448–529

Woman's side-laced ankle boot (prize work), Robert Dixon Box, England, about 1845
silk brocade & morocco; gold braid; leather; linen & silk lining

single straight; turned construction; square toe; stacked heel
15 x 226 mm
H4448–530

Woman's slip-on shoe (prize work), William Walsh for Robert Dixon Box, London, about 1851
leather & kid; gold fringe; leather
straight; turned construction; square rounded toe; stacked heel
18 x 225 mm
H4448–523

Page 46: **Woman's slip-on shoes, John Thomas for Henry Marshall, England, about 1885**
silk; leather; kid & linen lining
straight; turned construction; narrow oval toe; covered louis heel
paper label printed: By appt to her Majesty [Royal Arms]/ Henry Marshall, late Pattison, Bootmaker HRH Duke of Edinburgh, HRH Princess Louis of Hesse, HRH Princess Christian, 154 Oxford St London
79 x 243 mm
H4448–79

Page 47: **Woman's barette boots, Joseph Box, England, about 1896**
patent & glacé kid; leather; linen lining; jet beads
pointed toe; 13 straps buttoning at front; louis heel
woven: By appointment Joseph Box/ [the Royal Arms]/187 Regent St London
70 x 225 x 328 mm high
H4448–108

Page 49: **Woman's boots (prize work), Gundry & Sons, England, about 1880**
quilted silk; cork; leather; silk lining
rand construction; square toe; covered louis heel; 6 buttons
notes: made by Gundry & Sons, Soho Square, for the late Queen Alexandra when Princess of Wales
49 x 254 mm
H4448–517

Page 50: **Woman's tennis shoes (prize work), Joseph Box, England, about 1886**
morocco; rubber; kid lining
rand construction; narrow oval toe; spring wedge heel
stamped: [crown]/By Appointment/ Joseph Box/Regent St/London W
14 x 344 mm
H4448–536

Page 51: **Woman's oxfords (prize work), Joseph Box, England, about 1890**
morocco & glacé kid; leather; grosgrain ribbon; kid & linen lining
rand construction; pointed toe; louis heel
stamped:[crown] by appointment Joseph Box, Regent St, London W
71 x 254 mm
H4448–519

The 20th century to now

Page 52: **Woman's mules 'Luna', Camille Unglik (b 1940), Bally, Switzerland, summer 1970**
stainless steel; leather; rubber
instep strap; square shaped heel
stamped: 'Made in Switzerland'
56 x 235 mm
gift of Mary Shackman 1997
97/266/1

Page 55: **Woman's court shoes, Harrods, London, about 1910**
kid leather; glass; jet
pointed toe; beaded vamp; louis heel
stamped: Harrod's Ltd,/
Brompton Rd Sw
52 x 242 mm
purchased 1985
85/210–2

Page 56: from left
Woman's shoes, Pinet, Paris, about 1925
leather; mother of pearl
cross-over strap fastening; leather inlay pattern; louis heel
stamped: F. Pinet Paris/Exposition Universelle de M...CLXVII Paris
56 x 250 mm
purchased 1985
85/210–3

Woman's shoes, about 1925
silk brocade; leather decorative straps;
louis heel
53 x 230 mm
purchased 1985
85/210–8

Woman's shoes, Thierry, London, about 1925
crocodile skin
double strap; buckle fastening; louis heel
stamped: Thierry/Est 1839/
Regent Street, W1
60 x 240 mm
purchased 1985
85/210–4

Page 56: Man's shoes, Austria, about 1935
buckskin leather
welted construction; lace-up;
pinked edges
stamp: B. Nacy
28 x 300 mm
gift of Aviva Ziegler 1984
A10951

Page 57: Woman's shoes, Palter de Liso for David Jones, New York, about 1933
leather; satin; diamanté
t-bar; louis heel
stamped: David Jones/Sydney [left],
Palter De Liso Inc/New York City [right]
80 x 210 mm
purchased 1996
96/387/1

Page 58: Woman's shoes, Salvatore Ferragamo, Italy, about 1940
silk brocade; leather; suede; [cork]
sling back; oriental toe; wedge heel
stamped: Creations/Ferragamo's/
Florence/Italy
45 x 240 mm
purchased 1997
97/245/1

Page 59: Woman's sandals, Salvatore Ferragamo (1898–1960, established Florence 1929), Italy, about 1938
leather; [cork]; handpainted silk; glass beads
platform sole; padded straps
80 x 230 mm
gift of B C Reid 1976
A6624–58

Page 61: Woman's shoes, Roger Vivier (1913–98) for Christian Dior, Paris, about 1961
leather; silk
square toe; stiletto heel
stamped: Christian Dior/créé par/Roger Vivier/Paris/sur mesure
83 x 260 mm
purchased 1997
97/187/1

Woman's shoes, Raymond Massaro (b 1930), Paris, about 1965
leather; satin
sling back; needlepoint toe; heeled
stamped: Massaro/Opera 70–23/2,
Rue De La Paix. Paris
80 x 260 mm
gift of Betty Keep 1982
A8158

Page 62: from bottom to top
Woman's shoes, Roger Vivier for Christian Dior, about 1959
leather; nylon mesh; diamantés; beads
needlepoint toe; stiletto heel
85 x 240 mm
gift of Anne Fairbairn 1988 (TIAS)
88/842

Woman's shoes, Salvatore Ferragamo, Italy, about 1959
satin; leather; glass; diamantés
needlepoint toe; stiletto heel
stamped: Salvatore/Ferragamo/
Florence/Italy [left]; Made expressly/for/
David Jones [right]
82 x 240 mm
gift of Anne Fairbairn 1988 (TIAS)
88/840

Woman's shoes, Rayne (established London 1889), England, about 1959
leather; nylon mesh; embroidered cotton; diamantés
needlepoint toe; sides cut out;
stiletto heel
stamped: [right] Rayne
[left] Made In England/
David Jones/expressly for
82 x 235 mm
gift of Anne Fairbairn 1988 (TIAS)
88/841

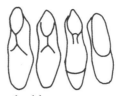

Page 63: from left
Man's boots ('chukka'), Harry B Hart, England, about 1955
suede; leather
ankle boots; lace-up; welted construction
26 x 285 mm
purchased 1984
A9932

Man's shoes, Alan McAfee (established 1848), England, about 1950–70
suede
lace-up; welted construction
stamped: [polo pony] 38 Dover St/Alan McAfee Ltd/London/W1/Real Cape Buck
23 x 285 mm
purchased 1984
A9931

Man's shoes, Alan McAfee, England, about 1940
leather
lace-up; toe cap
stamped: 38 Dover St/Alan McAfee Ltd/London/W1
23 x 280 mm
purchased 1984
A9930

Man's evening shoes, Alan McAfee, England, about 1970
patent leather, gros grain silk
evening pumps; bow decoration at upper
stamped: [polo pony] 38 Dover St/Alan Mcafee Ltd/London/W1
15 x 280 mm
purchased 1984
A9927

Page 64: Woman's mules 'Luna', Camille Unglik (b 1940), Bally, Switzerland, summer 1970
stainless steel; leather; rubber
instep strap; square shaped heel
stamped: Made in Switzerland
56 x 235 mm
gift of Mary Shackman 1997
97/266/1

Page 65: Woman's boots, Beth Levine (1914–2006), NY, about 1970
synthetic; plastic; perspex
thigh length; net stocking over perforated plastic shoe; rounded toe
stamped: herbert levine
33 x 240 x 615 mm high
purchased 1997
97/185/1

Page 66: Woman's shoes 'Kabuki', Beth Levine, NY, about 1964
leather; satin; lacquered wood
pointed toe; platform sole;
sculptured wedge heel
Label: Saks/fifth Avenue/Fenton Last/
patent pending
44 x 265 mm
purchased 1997
97/185/2

Page 66: Woman's ankle boots, Mary Quant (b 1934, established London 1963), England, 1960–1965
plastic; synthetic; metal; knit lining
moulded plastic construction; buckle fastening; daisy mould at heel
stamp: quant afoot
30 x 260 mm
purchased 1985
85/1231

Page 67: Woman's ankle boots, Beth Levine, NY, about 1970
leather; silk; satin; velvet
zipper fastening; patchwork decoration
stamped: Beth's Bootery [left]
Saks Fifth Avenue [right]
43 x 235 mm
purchased 1997
97/222/1

Page 67: Woman's clogs, Terry de Havilland, England, about 1974
leather; wood; metal
patchwork upper; nailed rim;
platform heel
110 x 230 mm
gift of A Young 1989
89/778/1:2

Page 68: Woman's shoes, 'Pensée' (Pansy), Christian Louboutin (b 1964, established Paris 1991), Paris, Autumn–Winter 1993–94
suede; silk; leather
pointed toe; flower decoration;
high-waisted stiletto heel
stamped: Christian/Louboutin/Paris/
Made in Italy
85 x 280 mm
purchased 1997
97/177/2

Woman's shoes, 'Mouse', Tokio Kumagai (1948–1987, born Japan to France 1980), Paris, Autumn–Winter 1985–86
leather; calf fur
slip-on; low wedge-heel
woven label: Tokio Kumagai/Paris/Made in Italy
35 x 285 mm
gift of L Simons 1995
95/11/1

Page 69: Woman's shoes 'Wannabe', Patrick Cox (b 1963 Canada, to Great Britain 1983, established London 1986)
leather
loafer style; stacked leather heel
woven label: Wannabe/by/Patrick Cox/
Made in Italy
60 x 270 mm
purchased 1995
95/31/1–1

Page 69: Man's boots, Dirk Bikkembergs (b 1962, born Germany), Belgium, about 1996
leather; metal; rubber; synthetic
ankle boots; front opening; wrap around straps; voided heel
woven label: Dirk Bikkembergs; stamped on sole: Made in Belgium
55 x 290 x 220 mm high
purchased 1997
97/136/1

Page 70: **Woman's shoes, Christian Lacroix (b 1951, established Paris 1987), Paris, Spring–Summer 1996**
reflective nylon; rubber
layered platform sole; lace-up
label: Bazar/de/Christian Lacroix
58 x 280 mm
gift of Christian Lacroix 1997
Rec. 7320

Page 71 : **Woman's shoes, 'Super Elevated Gillies', Vivienne Westwood (b 1941, established London 1980), England, designed Autumn–Winter 1993–94**
leather; cork; silk
concealed platform heel and sole; lace-up
stamped: Vivienne Westwood/London/ Made In England
210 x 240 mm
purchased 1997
97/208/1

Page 72: **Man's shoes, Nike, Taiwan, 1990–1999**
leather, rubber, polyester
inscribed inside: Made in Taiwan...990305 PC8
inscribed on sole: Nike Flight Systems Foamposite Technology
gift of Paul Jellard, 2006
2006/94/46

Page 73: **Woman's shoes, 'Splice Clogs', Royal Elastics, Australia, made in China, 1999**
leather, rubber
sole inscribed: Royal Elastics
105 x 165 mm
gift of Tull Price and Rodney Adler, Royal Elastics, 2001
2001/105/2–1

Page 74: **Man's sports shoes, 'Zvezdochka', Marc Newson (b 1963 Sydney), for Nike, Korea, 2004**
perforated rubber, plastic, synthetic mesh
0 x 285 mm
gift of Nike Australia Pty Ltd, 2005

Page 75: **Woman's boots, John Galliano (b 1960 Gibraltar) for Christian Dior, France, 2000**
denim, leather, metal
inscribed inside: Christian/Dior
inscribed on sole: Made in Italy / 39
85 x 244 x 500 mm high
gift of Christian Dior Australia Pty Ltd, 2000
2001/37/1–4/1:2

The Australian shoe industry

Page 76: **Man's oxfords, Donna-May Bolinger, Sydney, 1997**
crocodile skin; leather
hand-stitched welt construction
32 x 285 mm
inscribed: Handmade by Donna-May Bolinger
purchased 1997
97/237/1

Page 78: from left
Unfinished boot upper, Australia, about 1850
leather
235 mm long
gift of Frederic Thearle 1958
H5790–3

Miniature wellington boot, James Thearle, Australia, about 1839
kangaroo hide; leather
unfinished; stacked leather heel
28 x 155 x 260 mm high
gift of Frederic Thearle 1958
H5790–1

Miniature last, England/Australia, 1820–1839
wood; metal hinge
square toe; latchets
155 mm long
gift of Frederic Thearle 1958
H5790–2

Page 79: from left
Child's clogs, England/Australia, 1850–1870
wood; leather; metal fastenings; brass nails; iron sole rims
12 x 133 mm
purchased 1984
A10772

Child's clog-overshoes, England/ Australia, 1850–1870
wooden sole; leather straps; iron nails
33 x 146 mm
gift of C Davis 1958
H5540

Page 80: **Man's elastic-sided boots, R M Williams, South Australia, 1997**
leather; polyester bound core elastic; steel shank
one piece upper (trademark design); welt construction; stacked leather heel
woven label: R M Williams Made in Australia
30 x 290 x 170 mm high
gift of R M Williams 1997
97/219/1

Page 81: **Woman's evening sandals, Edward Meller, Sydney, 1940–1945**
leather; wood; cork
platform sole; open toe
stamped: Sydney/Meller Shoes/ 107A King St.
90 x 250 mm
gift of Lady Hurley 1978
A7094

Page 81: **Woman's surf shoes, Barnet Glass, Melbourne, Australia, 1920–1925**
rubber
moulded construction; instep strap
stamped: Barnet Glass
0 x 270 mm
gift of Sue McCredie 1996
96/342/1

Page 82: from left
Woman's shoes, Selby for David Jones, Australia, about 1935
suede; leather
oval toe; side-fastening buckle; louis heel
stamped: (left) David Jones/Limited/ Sydney (right) Selby/arch/preserver/ shoe/standard
65 x 235 mm
gift of J Lahm 1986
86/52

Woman's shoes, Westbrook for Farmers, Australia, about 1935
suede; leather; satin
lace-up; oval toe; louis heel
stamped: Model no 1311/Farmers/ Exclusive/[trade mark] Westbrook built
55 x 245 mm
gift of E Holliday 1985
85/81

Woman's shoes, The Public Benefit Bootery Ltd, Australia, about 1930
leather
latchet straps; oval toe; louis heel
stamped: PBB/Public Benefit/Bootery/ Pitt St Sydney
80 x 225 mm
purchased 1995
95/14/1

Page 83: **Man's winkle-picker shoes, Raoul Merton, Australia, about 1962**
leather
lace-up; tapered toe
stamped: raôul/merton/tailored shoes/ bonded construction/[plane]/from the 727 Series/numatic/cushion heel
25 x 350 mm
gift of Terence Mooney 1997
97/215/1

Page 84: **Woman's sandals, Merivale, Sydney, 1970–1978**
punched leather
platform sole; open toe; ankle strap
woven label: designed by Merivale
97 x 210 mm
gift of Ann Young 1989
89/777

Page 85: **Thongs, origin unknown, about 1980**
rubber; plastic
0 x 275 mm
gift of Martin Munz 1996
96/357

Page 87: **Man's shoes, 'Prince of autumn leaves', Pendragon Boot Co. (established 1987), Brisbane, 1994**
embossed leather; wood
layered leaf shapes; extended curled toe
inscribed: 'Pendragon/Boots: Jackie Orme'
30 x 310 mm
purchased 1997
97/178/1

Theatrical shoes, Jodie E Morrison (b 1953, established Sydney 1979), Sydney, 1997
leather
louis heel
woven label: Made especially for you by:/Steppin' Out/Sydney
58 x 260 mm
gift of Jodie E Morrison 1997
97/215/1

Page 88: **Woman's shoes 'Carita', designed by Philippe Modèle (b 1956, established Paris 1978) for Elle Effe, Sydney, 1989**
suede; leather lining
pump style
stamped: made by L F Footwear
28 x 235 mm
purchased 1997
97/233

Page 89 : **Woman's shoes, Donna-May Bolinger (b 1960 Canada, to Australia 1983, established Sydney 1987), Sydney, 1997**
screen printed leather
front fastening; straight heel
inscribed: Handmade by Donna-May Bolinger & maker's stamp
90 x 230 mm
purchased 1997
97/237/2

Page 94: **Woman's shoes, 'Little Woman Ware II', Gaza Bowen (b 1944), America, 1985**
sponge; Wisk detergent bottle; scouring pads; plastic scrubbies; linoleum; dish mop
purchased 1997
97/271

Back cover: **Woman's heelless shoe, Mario Brini, Italy, about 1959**
leather, steel
inscribed inside: Creazioni/Mario Brini/ brevetto/No 11233
75 x 230 mm
purchased 1997
97/246/1

With thanks

The Powerhouse Museum wishes to thank the following institutions, companies and individuals whose work is reproduced in *Stepping out: three centuries of shoes*.

Bally; Dirk Bikkembergs; Donna-May Bolinger; Gaza Bowen; Patrick Cox; Christian Dior; Terry de Havilland; Museo Salvatore Ferragamo; Leonie Furber; John Galliano; Christian Lacroix; Beth Levine; Christian Louboutin SARL; Alan McAfee; Raymond Massaro; Merivale Pty Ltd; Jodie Morrison; Marc Newson; Nike; Pendragon Boot Company; Mary Quant; Rayne; Royal Elastics; Camille Unglik; Vivienne Westwood; R M Williams.

Below: Californian artist Gaza Bowen (b 1944) uses the shoe as a medium for feminist commentary on women's role in society. These shoes made in 1996 entitled 'Little Woman Ware II' are made from sponges, a Wisk detergent bottle, scouring pads, plastic scrubbies, linoleum and a dish mop.

Further reading

Baynes, Ken and Kate (eds). *The shoe show: British shoes since 1790*, Crafts Council, London, 1979.

Bossan, Marie-Josephine. *The art of the shoe*, Parkstone International, New York, 2004

Delaney, Bill. 'From Erskineville to Ultimo: a history of the NSW School of Footwear 1906–1996', unpublished manuscript.

Devlin, James. *The guide to trade: the shoemaker*, Charles Knight, London, 1839.

Grew, Francis and de Neergaard, Margrethe. *Shoes and pattens: Medieval finds from excavations in London: 2*, Museum of London, 1988.

Heyraud, Bertrand. *5000 ans de chaussures*, Editions Parkstone, England, 1994.

Lehane, Brendan. *C & J Clark 1825–1975*, C & J Clark Ltd, Somerset, UK, 1975.

Maynard, Margaret. *Fashioned from penury: dress as cultural practice in Australia*, Cambridge University Press, Melbourne, 1994.

McDowell, Colin. *Shoes, fashion and fantasy*, Thames & Hudson, London, 1989.

Miller, Jerry. *The wandering shoe*, My Goodfriends, New York, 1984.

Probert, Christina. *Shoes in Vogue since 1910*, Thames & Hudson, London, 1981.

Provoyeur, Pierre. *Vivier*, Editions du Regard, Paris, 1991.

Riello, Giorgio & McNeil, Peter. *Shoes: a history from sandals to sneakers*, Berg Publishers, England, 2006

Swann, June. *Shoes*, Batsford, London, 1982.

Trasko, Mary. *Heavenly soles: extraordinary twentieth-century shoes*, Abbeville Press, New York, 1989.

Thearle, Susan. 'Thearles in Australia 1838 to 1988', unpublished manuscript, 1991.

Unorthodox Styles. *Sneakers: the complete collector's guide*, Thames & Hudson, London, 2005

Victoria & Albert Museum, *Salvatore Ferragamo: the art of the shoe 1927–1960*, Centro Di della Edifimi srl, Florence, 1987.

Weber, Paul. *Shoes: a pictorial commentary on the history of the shoe*, The Bally Shoe Museum, Schönenwerd, Switzerland, 1982.

Wilson, Eunice. *A history of shoe fashion*, Pitman, London, 1969.

About the authors

Louise Mitchell is a former curator of decorative arts and design at the Powerhouse Museum – responsible for the international fashion, textile and jewellery collections – and the curator of the exhibition *Stepping out: three centuries of shoes* (Powerhouse Museum, 1997–98). She has contributed to a variety of publications including *Decorative arts and design from the Powerhouse Museum* (Powerhouse Publishing, 1991); *Christian Dior: the magic of fashion* (Powerhouse Publishing, 1994) and *Heritage: the national women's art book* (an *Art and Australia* book, 1995). Her most recent exhibition was *The cutting edge: fashion from Japan* (Powerhouse Museum, 2005–06) and she edited the book of the same name.

Lindie Ward began her career as a fashion designer in London and Montreal and later specialised in historical costume for the theatre. Lindie works as a curator of design, history and society at the Powerhouse Museum, with responsibility for the textile and lace collection. She has contributed to a number of exhibitions and publications including *Bayagul: contemporary Indigenous communication* (2000); *Audrey Hepburn: a woman, the style* (1999); *Stepping out: three centuries of shoes* (1997); and *Christian Dior: the magic of fashion* (1994). As a complement to her curatorial work, she has created innovative ways to display and photograph costume both at the Powerhouse Museum and other galleries around Australia.

Christina Sumner is principal curator of design and society at the Powerhouse Museum. Specialising in textiles and textile technology, her research interests span the traditional cultures and textiles of Western, Central, South and Southeast Asia as well as those of Europe and Australia. She has curated numerous exhibitions at the Powerhouse Museum, and co-authored their associated publications, on the textile and other arts of the Asian region, most recently *Bright flowers: textiles and ceramics of Central Asia* in 2004, *Trade winds: arts of Southeast Asia* in 2001 and *Beyond the Silk Road: arts of Central Asia* in 1999.